Weapons & Field Gear
of the
North Vietnamese Army
and Viet Cong

Edward J. Emering

Schiffer Military/Aviation History
Atglen, PA

Dedication:

To the men and women of the Vietnam Veterans Art Group and in memory of our fallen brothers: David Sessions (1941-1994), Cleveland Wright (1931-1993) and Richard Yohnka (1951-1997).

Acknowledgments:

Special thanks to the National Vietnam Veterans Art Museum in Chicago for use of its facilities and collection and to Ned Broderick and Greg Goodrow for allowing me to photograph materials from their private collections and for their technical assistance. Thanks also to CDR. Frank C. Brown, M.D., USN (ret.), Gerry Schooler, Chuck Zimmaro and Tom Benedum for their assistance.

Book Design by Ian Robertson.

Copyright © 1998 by Edward J. Emering.
Library of Congress Catalog Number: 98-84113

Printed in China.
ISBN: 0-7643-0583-2

We are interested in hearing from authors with book ideas on related topics.

Published by Schiffer Publishing Ltd.
4880 Lower Valley Road
Atglen, PA 19310
Phone: (610) 593-1777
FAX: (610) 593-2002
E-mail: schifferbk@aol.com
Please write for a free catalog.
This book may be purchased from the publisher.
Please include $3.95 postage.
Try your bookstore first.

CONTENTS

PREFACE

From its borrowed and antiquated weapons, the People's Army of Vietnam (PAVN) has steadily improved and modernized its forces. The field gear used by PAVN (also referred to colloquially as NVA) and its National Liberation Front allies (referred to as the Viet Cong or VC), although often mismatched, but always utilitarian, is much sought after by collectors. The National Vietnam Veterans Art Museum (NVVAM), located in Chicago, Illinois, is no exception when it comes to collecting the field gear and medals of the "enemy." With a well established NVA/VC field gear collecting effort spearheaded by President and Founding Artist, Ned Broderick, and the medal collecting and cataloging efforts of Chairman and Photographer/Author, Ed Emering, the NVVAM has accumulated one of the finest and most extensive collections of its kind anywhere in the world. The tremendous diversity and range of weapons and gear in the NVVAM collection are showcased in this work.

It is generally held that NVA/VC field gear is divided into three periods: (1) pre-1958 when a Maoist inspired minimalist style was in vogue as reflected in the general lack of insignia usage; (2) 1958 through 1981, when the PAVN underwent a significant modernization program; and (3) post-1981 when major revisions in uniform regulations were implemented. Even with these major changes in uniform regulations, which occurred in 1982, as reflected in the increased formality of the PAVN's dress uniforms, one can still observe considerable diversity in the presence of battle ready PAVN forces. Many traditional pieces of field gear continue in use from the early years. Examples are the pith helmet, the bush hat and canvass combat shoe, which is now being exported to the former Soviet Bloc countries.

This work catalogs not only the advancements in weapons and field gear, but also the exceptionally broad scope of this collecting specialty. For Ed Emering and his primary technical advisors, Ned Broderick and Greg Goodrow, this has been a labor of joy, which is reflected in the extensive nature and caliber of this work.

C. N. Bergold
Chicago, Illinois

INTRODUCTION

From its humble beginnings in the early 1940s as a 34 man "praetorian guard" for Ho Chi Minh, equipped with rudimentary weapons, including reportedly flintlock rifles in some cases, the Viet Minh, as they were known, and their successors, the People's Army of Vietnam (PAVN), have grown into one of the largest and most battle hardened armies in the world. It is second only to the Communist Chinese (ChiCom) Army in terms of size. Under the command of the famous Senior General Vo Nguyen Giap, the PAVN initially relied on "borrowed" World War II U.S. and Chinese weapons and field gear in the beginning. This trend would continue through its three major wars, fought between 1950 and 1985. The PAVN, which was known for its self sufficiency in so many other areas, would rely heavily on Russian and ChiCom weaponry and gear as it grew into a major Southeast Asian military force. Even during its second major war against the South Vietnamese and their American allies, very few pieces of gear were actually manufactured in North Vietnam. For the most part even military training was provided by Russian and ChiCom schools and advisors.

PAVN 1944 to 1954

The focus of PAVN's initial concern was the invading WW II Japanese Army. The active presence of the Imperial Japanese Army in southern China drove the early PAVN forces and their leaders to seek refuge deep within the rugged sanctuaries of Northwest Vietnam's Cao Bang Province, located only one mile south of the Chinese border. It was there in February of 1941, that Ho Chi Minh established his headquarters in the Pac Bo Cave in the District of Ha Quang in Cao Bang Province. There, PAVN marks the date of its official beginning as December 22, 1944. By the end of 1944, the PAVN had more than 10,000 troops. The Pac Bo Cave would continue to serve as a Communist sanctuary throughout the Second Indochina War.

With the defeat of the Japanese by the allied forces, Ho Chi Minh and his Viet Minh followers attempted to seize power. In September 1945, with the PAVN now numbering more than 30,000, Ho Chi Minh read his declaration of independence in Hanoi, proclaiming the establishment of the Democratic Republic of Vietnam. This independence would be short lived. A much stronger, better equipped French (Foreign Legion) Expeditionary Force would push Ho and his followers out of the major cities and back into the moun-

tainous lair of Cao Bang Province. By 1947, when the Viet Minh went to war against the French, the PAVN had grown to more than 100,000 strong. A protracted war against the French would be launched on two levels. On one level, main force units, such as the "308 Capital Division" composed mainly of ethnic Vietnamese from the Hanoi region, would hold their own against the French units. On the other, militia or guerrilla units were used to harass the French at the village and hamlet level. By 1950, PAVN's strength at 300,000 would rival that of the French with the only exception being the French air capabilities. With ChiCom training, bolstered by ChiCom

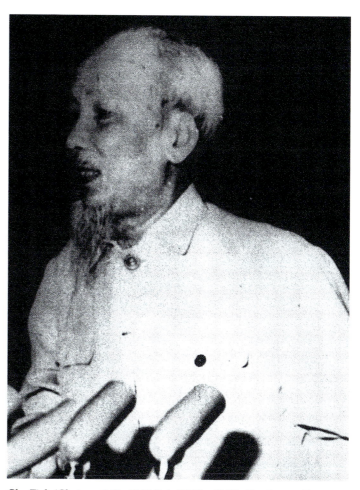

Chu Tich (Chairman) Ho Chi Minh (1890 - 1969). Goodrow collection.

field advisors, the PAVN successfully prosecuted their war against the French along the Sino-Viet border area.

These successes would eventually set the stage for the 1954 showdown between the French forces and General Giap at Dien Bien Phu where the resilient Viet Minh forces would surprise the entire world. More than 200 heavy artillery pieces, including 75mm, 105mm howitzers and Soviet rocket launchers, would be dismantled and hauled into impregnable mountain side emplacements overlooking the French positions in the valley. Giap would have at his disposal more than 100,000 artillery shells, 3,000 various rifles and 2.4 million rounds of ammunition. At a cost of more than 500,000 men killed in action during the course of the war, General Giap's PAVN forces would bitterly defeat the French on May 8, 1954. As a result, Vietnam would be divided into two independent countries at the 17th parallel under the terms of the Geneva Accords. The PAVN would use this period to prepare itself for the ultimate unification of the country, which would take longer than two decades to accomplish. General Giap would emerge as one of the world's great military strategists and his lieutenant, Colonel General Chu Van Tan, a Nung tribesman, as a great field commander.

PAVN 1955 to 1975

Following the First Indochina War, PAVN undertook a major modernization program. Supported by both Russian and ChiCom training and equipment, the PAVN made major strides forward, introducing both a naval and an air element to its overall force strength. Russian and ChiCom advisors, stationed in North Vietnam, helped to supplement the training being received in China and the Soviet Union.

As early as 1959, PAVN forces began heading south across the 17th parallel. The highly secret 559th Military Transportation Group, commanded by Major General (then Senior Colonel) Vo Bam, was established on May 5, 1959 to organize and maintain a logistical supply route into South Vietnam. This supply route for men and material, originally called the Truong Son Route, would become known to the world as the Ho Chi Minh Trail. In July of the same year studies were undertaken, which would lead to the activation of the 759th Military Transportation Group on October 23, 1961. Group 759, commanded by Doan Hoang Phuoc, was charged with opening a sea infiltration route into South Vietnam. In 1963 Group 759 was redesignated Group 125. The goal of this infiltration was to support the growing guerrilla effort of the National Liberation Front (NLF, or VC as used herein). This guerrilla activity was placed under the control of the Central Office for South Vietnam (COSVN), which took its direction from the Politburo in Hanoi. With PAVN support, COSVN also operated on two levels, using both the main force units of the People's Liberation Armed Force (PLAF) starting in 1961 as well as local guerrilla units. By 1963, using PAVN leadership and Russian and Chinese supplies ferried down the Ho Chi Minh Trail, the NLF had formed two PLAF regiments of its own. These PLAF forces would shoulder the major burden of the war through the 1968 Tet Offensive. NLF losses during the 1968 Offensive were so extensive that PAVN forces would be required to lead and staff the NLF units for the balance of the Second

Indochina War. During the next seven years, PAVN would launch two major offensives in the South; the first in 1972 (the Nguyen Hue Offensive), relied extensively on Soviet and Chinese armor. More than 14 divisions invaded the South, but after initial successes, the Offensive proved to be a major fiasco. The Army of the Republic of Vietnam (ARVN), supported by U.S. air power, enjoyed its finest moment.

More than 600 PAVN armored vehicles were destroyed and nearly 200,000 PAVN soldiers killed. The invasion was repulsed with no permanent loss of territory. PAVN would lick its wounds until spring 1975. It is interesting to note that throughout its history, whenever the PAVN suffered a major defeat, it seemed to learn its lesson well. PAVN did not often repeat the same mistakes.

The intervening period saw the signing of the Paris Agreements and the withdrawal of U.S. forces in 1973, eliminating a potent obstacle to PAVN's objective.

Without the intervention of U.S. air power, PAVN's forces, now under the field command of Senior General Van Tien Dung, enjoyed rapid and surprising success. After initial defeats in I Corps (the northern most theater of operations in South Vietnam) and the abandonment of the Central Highlands by ARVN forces, the PAVN rolled into the coastal zones to the north and east of Saigon. After some resistance at Xuan Loc, PAVN armored forces in T-54 tanks, flying the NLF flag, crashed through the gates of the Presidential Palace in Saigon on April 30, 1975 bringing the Second Indochina War to an end.

PAVN 1976 to Present

With the end of the protracted war of unification, PAVN turned its focus to implementing the Politburo's policies in the South and establishing control over the region. During this same period, the People's Liberation Armed Forces, numbering approximately 300,000, was abolished as a separate entity. Most of the NLF main force cadre, which were PAVN fillers to begin with, were absorbed into other PAVN units.

Almost simultaneously (1975), Pol Pot's Khmer Rouge (Democratic Republic of Kampuchea) began a series of small skirmishes along the Cambodian border areas. As these continued, the Politburo lost patience with the radical Khmer. On Christmas Day 1978, it ordered a full scale (Russian Style) invasion of Cambodia under the leadership of Senior General Van Tien Dung. It was widely thought that this incursion would be brief and productive. Hanoi expressed much optimism for an early conclusion to this operation when PAVN units occupied Phnom Penh on January 7, 1979. Their optimism was short-lived though as the six PAVN divisions and two marine units (150,000 men), bolstered by the assault youth forces, (Thanh-Nien Xung-Phong or TNXP), became mired in a protracted guerrilla war of their own (their third full scale war of this half century). The Khmer Rouge had taken refuge in the rugged Cardamom mountains where they would continue to strenuously resist the PAVN invaders (even to this very day) with material and financial support from Communist China.

PAVN's invasion of Cambodia greatly displeased their Chinese benefactors, who had also supported Pol Pot. On February 17,

1979, following a heavy artillery bombardment along the 480 mile border with Vietnam, 85,000 men from the Chinese People's Liberation Army invaded Vietnam on a punitive mission. The Chinese forces attacked Lao Cai, Cao Bang and Lang Son, placing the invaders less than 100 miles from Hanoi. PAVN countered with 100,000 men of their own. After several days, the Chinese advance was halted and a classic artillery duel ensued with shelling the heaviest around the town of Lang Son, which eventually fell to the Chinese on March 4. The very next day, with the realization that it would be very costly to attempt to occupy the overrun territories, China announced that the "lesson" had been taught and that they were withdrawing their forces from Vietnam. PAVN lost an estimated 35,000 combatants during the brief 17 day phase of the overall Sino-Vietnam War; the PLA an estimated 30,000 soldiers. Senior General Dung linked the Kampuchean Conflict directly to the ongoing disagreements with China. Military tension with China would continue to fester. It would flare again in March of 1988, when the ChiCom Navy would sink two Vietnamese supply ships in the waters around the disputed Spratly Islands.

By 1979, PAVN controlled most of the major Cambodian population centers, but the Khmer Rouge still held sway in the countryside as numerous PAVN efforts to uproot them would end in failure and frustration. PAVN occupation of Cambodia would continue for another decade into 1989 and eventually, the responsibility for continued prosecution of the war would be turned over to the Army of the People's Republic of Kampuchea (PRK), the Vietnam puppet state. Vietnam's military presence in Cambodia continues to linger in the form of advisors, special operations (Dac Cong commando) units, and fillers in the PRK Army on much the same basis as fillers had been provided for the PLAF in an earlier war. PAVN has faced many of the same challenges in relation to the People's Republic of Kampuchea (PRK) Army as the Americans experienced with regard to the ARVN during their Southeast Asian War. The PRK Army seems content to protect the major population centers, but exhibits little interest in driving the Khmer Rouge from their countryside enclaves.

During this entire time, the PAVN also maintained about 30,000 men in Laos to help bolster and support the Communist government against insurgent anti-Communist forces. PAVN has also had to deal with a series of internal and external resistance movements during this same period of time. Resistance has stemmed from such diverse sources as religious sects, ethnic minorities, former ARVN

Senior General Vo Nguyen Giap circa mid 1950s. General Giap is wearing the Military Exploit Order on an early straight suspension ribbon and the first version of the Ho Chi Minh Order. He would later be awarded a second Ho Chi Minh Order and on August 20, 1991, he would receive the SRV's highest decoration, the Gold Star Order. His uniform shirt is most likely of ChiCom origin. U.S. Army Photo.

personnel and former PLAF personnel. PAVN has also had to wrestle with serious economic woes, which beset the country. Regardless, PAVN seems resilient and able to cope well with adversity. Today, PAVN's forces number in excess of 1 million with approximately an additional 2.5 million men in reserve. No doubt exists over the fact that PAVN is a force to be reckoned with in Southeast Asia.

GLOSSARY OF TERMS

ARVN - acronym for the Army of the Republic of Vietnam.

ChiCom - abbreviation for Chinese Communist.

DRK - acronym for the Democratic Republic of Kampuchea headed by Pol Pot and the murderous Khmer Rouge.

DRV - acronym for the Democratic Republic of Vietnam, predecessor of today's Socialist Republic of Vietnam.

HE - acronym for High Explosive.

HEAT - acronym for High Explosive Antitank.

Howitzer - a cannon with a comparatively short barrel.

Kampuchea - Cambodia.

Khmer Rouge - radical communist movement based in Cambodia.

MACV-SOG - acronym for the Military Assistance Command Vietnam - Studies and Observation Group.

NCO - acronym for Non-Commissioned Officer, i.e. generally holding the rank of Corporal or above, but not a Commissioned Officer.

NLF - acronym for the National Liberation Front.

NVA - acronym and the American term for the North Vietnamese Army (PAVN).

PAVN - acronym for the People's Army of Vietnam (Quan Doi Nhan Dan).

PLA - acronym for the People's Liberation Army of Communist China.

PLAF - acronym for the People's Liberation Armed Forces (Viet Cong).

PRK - acronym for the People's Republic of Kampuchea (Hanoi's client state).

SRV - acronym for the Socialist Republic of Vietnam, successor to the DRV.

Viet Cong (VC) - the South Vietnamese and American slang term (and acronym) for the soldiers and guerrilla fighters of the National Liberation Front. It is a shortened version of Viet Nam Cong San.

Viet Minh - the resistance front against the French Colonial forces and the forerunner of today's PAVN.

1

LIGHT INFANTRY WEAPONS

As previously mentioned, from its humble beginnings in the 1940's, the PAVN has emerged as one of the most experienced armies in the world. The backbone of any fighting force is the weapons it uses. From a hodge podge of various weapons captured and/or donated to the Viet Minh forces in the 1940s, the PAVN, and ultimately the post-1968 PAVN staffed and led Viet Cong units, relied heavily on the Soviet Union and China as primary sources of infantry weapons.

Pistols

An extensive array of pistols were used in the Second Indochina War, including the French 7.65mm M1935-A, the Czech 7.65mm M1950, the German 9mm Walther P-38, the Japanese 8mm Nambu Model 14 and the Soviet 9mm Makarov. The standard pistol, however was the Soviet Tokarev and its ChiCom copies Types 51 and 54.

Tokarev - The standard issue pistol was the eight round Soviet 7.62mm x 25 Tokarev (TT-M1933) automatic and its ChiCom copies, Type 51 and Type 54. These weapons were carried in the ChiCom Type 54 holster. The pistol has no safety device. These were generally issued to officers only, but on occasion to senior NCOs also. In combat, officers, particularly NLF officers, were often identified by the presence of a pistol.

Regardless of the extent of the array of available manufactured pistols, many used were also of the "homemade" variety, especially those used by the NLF or Viet Cong. Skilled guerrillas were very capable of manufacturing copies of existing weapons in crude jungle workshops when specimens of the originals, such as the Browning GP35 and the Walther P38 9mm pistols, were available to them.

ChiCom version (Type 51) of the eight round 7.62mm Soviet Tokarev (TT-M1933) pistol and ChiCom Type 54 holster with two external cartridge pouches.

Homemade Viet Cong pistol. Navy Museum, Washington, D.C.

Homemade Viet Cong pistol. Navy Museum, Washington, D.C.

Homemade Viet Cong pistol. Navy Museum, Washington, D.C.

Homemade Viet Cong pistol.

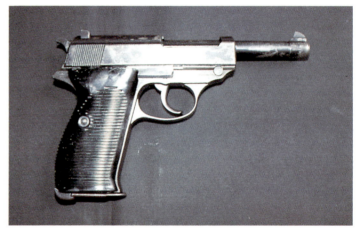

Viet Cong copy of the German Walther P38 9mm parabellum pistol.

ChiCom copy of the Browning GP35 9mm parabellum automatic pistol. Manufactured in large numbers by Fabrique Nationale of Belgium, the pistol was used extensively in China during World War II and was most likely copied by the Chinese at that time.

Homemade Viet Cong 12 gauge shotgun. The buttstock is made from salvaged aircraft parts.

Homemade Viet Cong rifle.

Close-up of the buttstock of the homemade rifle. "Gia Dinh To Quoc" translates as "Family and Fatherland."

Submachine Guns

PPSh 41 - Among the first Russian submachine weapons introduced into PAVN units were the PPSh 41 (and its ChiCom version Type 50), designed by Georgi Shpagin, and the PPS 43, designed by Sudarev. The PPSh 41, made from steel pressings and a wooden stock, was most frequently configured with a 71 shot drum magazine. The weapon is also encountered with a 35 round stick magazine. Russia produced more than five million of these weapons and they later became a mainstay of the PAVN Border and Coastal Defense forces during the 1940s and 1950s. Their design was copied by the Chinese, who called their version the Type 50 submachine gun. The PPSh 41 is capable of firing 90 to 100 rounds per minute on full automatic.

PPS-43 - The PPS 43 is a crudely made, but highly effective blow back submachine gun. Formed from spot welded metal pressings, the PPS 43 features a metal folding stock and a 35 round, curved box magazine. The design was copied by the Chinese after World War II and given the nomenclature Type 43. The two models are practically indistinguishable, except for the Chinese markings on top of the receiver and the diamond design on the grip. This gun is capable of firing 100 rounds per minute.

MAT-49 - A third early version of the submachine gun, utilized by enemy forces in Vietnam, was the captured French MAT-49. The blow back, 9mm MAT was a descendant of the pre-World War II MAS 38. It was converted to accept the Soviet (and ChiCom) 7.62mm x 25 pistol cartridge by modifying and elongating the barrel. The modified weapon used the nomenclature MAT-49-Mod. It was employed extensively by the Viet Cong during the Second Indochina War. It is easily recognized by its round perforated barrel jacket, prominent grip safety, and 32 round box magazine. It is capable of firing 120 rounds per minute.

Kar 98K ("Kurz") bolt action 7.92mm Mauser rifle. This was the standard rifle of the German Army during World War II and used by the French Colonial forces during L'Guerre de Indochine. Captured French pieces as well as those manufactured in East Germany were funneled to the Viet Cong during the Second Indochina War.

PPSh 41 assault rifle configured with a 35 round stick magazine.

Viet Minh officer (1950s) with PPSh 41 and four pocket khaki canvas chest pouch with leather flaps for the 35 round stick magazine. The metallic whistle worn around his neck identifies this man as an officer. Emering collection.

PPS-43 assault rifle.

7.62mm MAT-49 Mod submachine gun (SMG).

Viet Cong guerrilla with Nationalist Chinese Type 36 copy of the American .45 caliber M3A1 blowback "grease gun." These guns were originally manufactured as 11mm weapons, but were converted to accept the 9mm parabellum round. Emering collection.

ChiCom Type 56 copy of the Soviet SKS 7.62mm assault rifle.

Close-up of ChiCom Type 56 SKS assault rifle. Note Chinese characters on the stock.

Assault Rifles

SKS - One of the original self-loading assault weapons introduced into PAVN units was the gas operated, 7.62mm Russian SKS. Developed during World War II, the SKS featured a 10 round magazine and a fixed bayonet folded under the barrel. The SKS became a popular Russian export, used to supply various communist inspired guerrilla organizations around the world. The SKS (ChiCom Type 56) was also made in China and exported to North Vietnam. Later versions of the ChiCom copy featured a folding spike bayonet as opposed to the flat blade bayonet. The rifle could fire 30 to 35 rounds per minute.

Moisin-Nagant Carbine - The shortened ChiCom Type 53 carbine is a copy of the 7.62mm Soviet Moisin-Nagant M 1944 carbine. It is a descendant of the Soviet M1891/30 rifle. It has a manual bolt action and was also encountered with a spigot type grenade launcher. In the grenade launching configuration, it was referred to as the "Red Stock" or AT-44. It utilized a five round magazine. As the AT-44, it was equipped with high explosive (HE) rifle grenades with 22mm diameter tail booms and 7.62 x 54 rifle grenade launching cartridges. It was generally considered obsolete for use by PAVN

and main force VC units by the early 60s, but was often encountered in skirmishes with local VC guerrilla units. It could fire 10 rounds per minute.

AK-47 - The most popular and highly prized enemy weapon of the Second Indochina War was the gas, selective fire AK-47 assault rifle. Based on the German Sturmgewehr, designed by Louis Schmeisser, the AK-47 became the Russian assault rifle of choice. Designed by Mikhail Kalashnikov, it fires a 7.62mm cartridge from a 30 round curved "banana" clip magazine and has developed a

Soviet 7.62mm Ak-47 assault rifle.

ChiCom carbine Type 53 copy of the Soviet Moisin Nagant manual bolt action rifle.

Close-up of the selector switch on the Soviet Ak-47 assault rifle.

ChiCom Type 56 (AK-47 copy) assault rifle.

Close-up of the selector switch on the ChiCom Type 56 assault rifle.

PAVN (wearing sun helmets) and NLF (wearing soft boonie hats) soldiers circa early 1960s. Officer (left) is identified by the "hero" pens in his left breast pocket. These pens were bestowed on soldiers as an emulation badge and read "hero" in English. PAVN soldier (second from left) carries the ChiCom Type 56 assault rifle. NLF soldiers (third from left and far right) carry the ChiCom Type 56-1 folding stock assault rifle and a captured U.S. M-79 grenade launcher, respectively. Note that all are wearing Ho Chi Minh sandals. Emering collection.

well earned reputation for its reliability and ruggedness. It was manufactured with both a wooden butt stock as well as a metal folding stock. The Chinese assault rifle Type 56, which was manufactured with a wooden stock, is a direct copy of the AK-47 and is recognizable by its permanently affixed bayonet under the barrel. The Chinese also produce a metal folding stock variation, known as the Type 56-1. Both the Russian and ChiCom versions saw extensive action in Vietnam and have replaced the SKS in terms of popularity with Communist guerrilla organizations throughout the globe. It is capable of firing 90 to 100 rounds per minute.

Portable Rocket Launchers

RPG-2 - The Soviet RPG-2 (ChiCom antitank grenade launcher Type 56) is a muzzle loaded, shoulder fired, smooth bore, recoiless launcher, which fires a fin stabilized 40mm round. A gas escape port is located on the right side of the weapon making it impossible to fire from the left shoulder. It weighs a mere 6.3 pounds and can fire four to six rounds per minute. The North Vietnamese/VC designated this weapon as B-40. They also used an elongated 50mm copy with wooden tube insulation, known as the B-50. The B-50

RPG-2 portable rocket launcher.

version can be either shoulder fired or ground fired using an adjustable monopod.

RPG-7 - This is also a recoiless, shoulder fired, muzzle loaded antitank grenade launcher. It was modeled after the RPG-2, but with many improvements. The larger tube fires an 85mm round with much more thrust than the RPG-2. It also has a detachable telescopic sight. The weapon weighs 14.5 pounds. There are several variants of the RPG-7, in particular the V and D variants.

NLF main force soldier with RPG-2 portable rocket launcher, which is shoulder fired from the right side only. Emering collection.

RPG-7 portable rocket launcher.

RPG-7 optical sight.

RPG-7V.

RPG-7V optical sight.

ChiCom Type 53 copy of the Soviet DPM light machine gun (LMG).

ChiCom Type 56 copy of the Soviet RPD LMG.

Two NLF main force soldiers: the soldier (right) carries the ChiCom Type 56 assault rifle with wooden buttstock while the other (left) carries a ChiCom copy of the Czech ZB-26 light machine gun. Note the variation in uniform shirt colors. Emering collection.

ChiCom "Gimo" copy of the Czech Model ZB-26 LMG.

MG-34 LMG, which was most likely captured from the French during the First Indochina War.

Light Machine Guns (LMG)

A wide variety of light machine guns were used by the PAVN and the VC, ranging from captured French weapons to highly utilitarian Soviet and ChiCom weapons.

Soviet DP and DPM - The gas operated, air cooled Degtyaryev, designed by Vassily Degtyaryev, first saw action as early as 1928. It uses a flat drum 49 shot magazine to fire its 7.62mm round. It is reported that accuracy was improved by only loading 47 rounds into the magazine. Many difficulties encountered with the original DP model were overcome by slight design modifications in the model DPM, introduced in 1945. Both the DP and DPM models, as well as their ChiCom Type 53 copies, were used by the PAVN and Viet Cong. It is capable of firing 80 rounds per minute and weighs justly slightly more than 26 pounds.

Soviet RPD - The RPD was the last Russian Degtyareyv designed machine gun. It utilized the same 7.62mm intermediate round as the AK-47 assault rifle. Considered highly reliable, the RPD is gas operated and weighs only 19.4 pounds. The ChiCom copy is known as the Type 56. A North Korean copy is referred to as the Type 62. The RPD eventually became the PAVN's standard light machine gun during the Second Indochina War. It is capable of firing 150 rounds per minute. An extensively modified version is known as the RPDM or Type 56-1 and is considered even more reliable than the older version.

Czech Model 26 - The Chinese also furnished the PAVN with their Gimo copy, Model 26, of the gas operated, air cooled Czecho-slovakian 7.92mm ZB 26 light machine gun, originally designed by the firm of Zbrojovka Brno. The gun utilizes a distinctive top fed curved magazine. The gun weighs less than 22 pounds and is capable of firing 180 to 200 rounds per minute.

German MG 34 - The air-cooled, belt-fed (five 50 round metallic link belts) MG34 had its origins in 1934 Germany. Many models have a double trigger mechanism allowing for selective rates of fire. It was used extensively by the French Colonial forces in Indochina and large quantities of the weapon were captured by the North Vietnamese and placed in use during the Second Indochina War. Others were provided to the PAVN by the East German Army, which still used the MG-34 during the Second Indochina War period. It is operated by a combination of gas and recoil. It can also be used with either a 50 or 75 round drum magazine fitted into the top of the receiver. Extremely well made, this 7.92mm light machine gun often fell victim to dirt and subsequent fouling. It weighs less than 27 pounds and is capable of firing 100 to 120 rounds per minute. It is also capable of being adapted for antiaircraft use.

French M1924/M29 - The 7.5mm, gas operated, air cooled Chatellerault M1924 was developed by the French National Arsenal. It feeds from a straight magazine mounted above the receiver and it is capable of selective rates of fire. It was used extensively by the French Foreign Legion forces in their Colonial Wars, including L'Guerre de Indochine. Large, captured quantities of Type M29 of this machine gun were used extensively by the Viet Cong. It weighs 24 pounds and fires at an automatic rate of 125 rounds per minute.

French 7.5mm Chatellerault Type M29 LMG. Large quantities of these guns were captured from the French by the Viet Minh and subsequently passed along to the Viet Cong during the Second Indochina War.

2

CREW SERVED WEAPONS

Heavy Machine Guns (HMG)

Maxim - The water cooled Maxim was first introduced in Germany in 1887. Modified to the model MG08, it became the backbone of the German army during World War I. The Soviet 7.62 mm version (M1910 SPM) dates from 1910. The modified, 7.92mm Type 24 (1935), manufactured in China prior to World War II, was supplied to the PAVN. It is an extremely heavy weapon, weighing more than 50 pounds, with the mounting device weighing an additional 70 pounds. It is capable of firing at a rate of 200 to 300 rounds per minute. It was often employed in an air defense role.

PAVN soldiers pose in front of a ChiCom Type 54 (1935) Maxim HMG. Emering collection.

ChiCom Type 54 copy of the Maxim MG-08 7.92mm heavy machine gun (HMG).

Soviet Goryunov 7.62mm HMG, which was used extensively by PAVN units during the Second Indochina War.

Soviet DShK 12.7mm HMG, which was used extensively by PAVN primarily in an antiaircraft mode. It utilizes a mechanical computer sighting device.

Close-up of the mechanical computer sighting device mounted on the DShK HMG.

Czech T-21(Tarasnice) 82mm recoiless antitank gun.

Soviet SPG-82 rocket launcher, which was used primarily as a light weight infantry antitank weapon.

Soviet SG-43/SGM - The Goryunov M1943 (SG) Soviet (ChiCom Types 53 and 57) 7.62mm, gas operated, air cooled machine gun was the standard Communist heavy machine gun during the Second Indochina War. These guns can either be wheel mounted or vehicle mounted. The SGM version has a fluted barrel and is capable of firing 250 rounds per minute. The SGM weighs nearly 30 pounds and its wheeled mount adds another 50 pounds.

Soviet DShK 38/46 -The 12.7-mm Soviet DShK 38/46 (ChiCom Type 54 HMG) is a gas operated, air cooled versatile heavy machine gun, which was used primarily as an antiaircraft weapon in North Vietnam during the Second Indochina War. The gun weighs nearly 63 pounds and the mount, depending on variety, can add to this total substantially. The model 38/46 features a shuttle feeder with a flat feed cover. It is capable of firing 80 rounds per minute with an effective antiaircraft range of 1,000 meters. It uses a mechanical computer as an antiaircraft sighting device.

Anti-Tank Guns, Rocket Launchers and Mortars

Czech T-21 - The Czechoslovakian, 82-mm, Type 21 (Tarasnice), recoiless gun was used by both the PAVN and the VC as a lightweight infantry anti-tank weapon. Its smooth bore fires a fin stabilized HEAT projectile (shaped charge), ignited by a trigger activated magneto. It can be fired from a small two wheel carriage, from the shoulder or from a fixed mount. It weighs approximately 44 pounds and can fire four to six rounds per minute.

SPG82 Rocket Launcher - This infantry rocket launcher, which was already obsolete in the Soviet Union, was used primarily by the PAVN as an antitank weapon during the early days of the Second Indochina War. It is usually fired from its two wheel mount and can fire either an 82mm HE or HEAT round from its 84 inch barrel with a range of approximately 275 to 300 meters. The integral shield is designed to protect those firing the weapon from its heat blast. It weighs 83 pounds.

NLF air defense unit trains for antiaircraft operations under the watchful eye of an officer with a captured U.S. 50 caliber heavy barrel version M2 HMG mounted on a DShk 38/46 tripod. Also note the use of the manual computer usually mounted on the Soviet DShK HMG. Goodrow collection.

ChiCom 87mm Rocket Launcher - The Type 51 rocket launcher, first produced in China during the Korean Conflict, is a direct copy of the U.S. 3.5 inch rocket launcher M20. The ChiCom tube is of one piece construction compared to the American weapon's two piece aluminum tube. The ChiCom weapon also has a flared muzzle flash deflector, which is perforated with one-quarter inch holes on the right side. The smooth bore, breech-loaded, electrically fired weapon is also fitted with a forward-folding bipod. The weapon weighs 21 pounds, is five feet long and can fire two to four armor piercing shells per minute.

Soviet 122mm Rocket Launcher - Used extensively by the PAVN during the Second Indochina War, this single shot rocket

Close-up of the sighting optics on the Soviet 122mm single shot rocket launcher.

Soviet 122mm single shot rocket launcher and its 76 inch high explosive (HE) rocket below it.

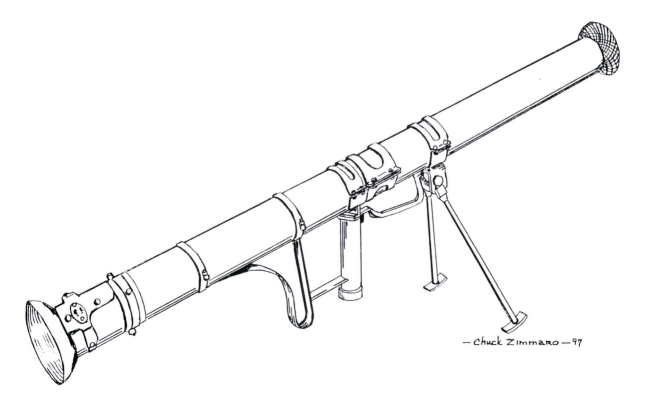

ChiCom Type 51 87mm HEAT rocket launcher. Artist C. P. Zimmaro.

Soviet 50mm mortar.

launcher consists of a 96.2 inch launching tube, a tripod mount and a panoramic sight. It fires the Soviet 122mm HE, fin stabilized rocket. Its serrated steel sleeves break into diamond shaped fragments when the warhead explodes. The PAVN referred to this system as the DKZ-B. The launching tube weighs 48 pounds and the tripod an additional 61 pounds. The rocket is 75.4 inches long and weighs just slightly more than 100 pounds. The weapon had a maximum range of nearly 11,000 meters.

50mm Mortar - Developed by the Russians in the 1930s, this variable gas system mortar has a 31 inch barrel. It weighs only 25 pounds and has an effective range of 800 meters. There are four models of this weapon: the M38, 39, 40 and 41.

60mm Mortar - This weapon was manufactured by the French (M1935), the U.S. (M2) and the ChiCom (Type 31). The model is recognizable by the square base and hand crank at the end of the elevating screw housing. The tube of the Type 31 is just slightly more than 26 inches long and the weapon weighs almost 45 pounds. It has a range of 1,530 meters.

ChiCom/NVA 60mm Knee Mortar - This an adaptation of the WW II Japanese Knee Mortar. The NVA model fires conventional 60mm fin stabilized shells at a variable elevation. The mortar is fired with its base plate held firmly against the earth between the

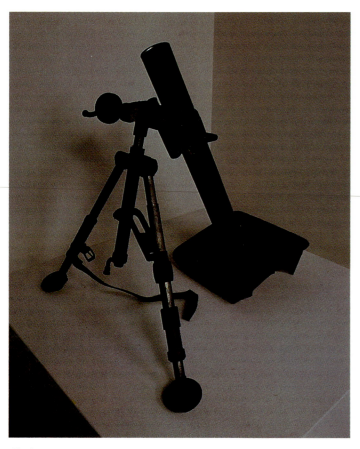

ChiCom Type 31 60mm mortar with distinctive square base plate.

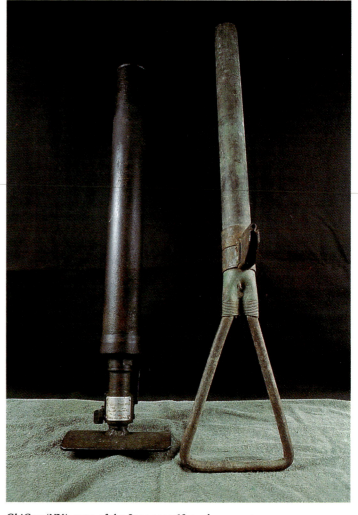

ChiCom/NVA copy of the Japanese 60mm knee mortar.

knees of the shooter. It has a base plate, which reads: "Giai Phong" (Liberation).

82mm Mortar - The Soviet 82mm mortar M1937 mortar (Chinese Type 53) is a conventional smooth bore, drop fire weapon. It is easily recognized by its circular base plate and its turn buckle cross leveling device. The tube is 48 inches long. It has an effective range of 3,040 meters and weighs 123 pounds.

Soviet 140mm Rocket & Launcher - This rocket marks the height of simplicity. The rocket is fired from a 45 inch metal cylinder (usually taken from the Soviet RM141 launcher) mounted to a wooden plank, which was part of the original shipping container. Significant adjustments in elevation are accomplished by piling dirt under the muzzle of the launching tube. The rocket is 43 inches long and weighs approximately 85 pounds. It is spin stabilized by ten exhaust nozzles in the base section. The fuse is point detonating with either short or long delay.

Close-up of base plate on the NVA 60mm knee mortar, which reads "Giai Phong" (Liberation). The presence of this plate on captured mortars indicated that the tube was manufactured in North Vietnam, however the tube was most likely manufactured in a ChiCom factory and assembled in North Vietnam where the plate was affixed for morale reasons.

Soviet 82mm mortar.

Soviet 140mm launcher and rocket. The launcher is secured to a wooden plank, which was part of the original shipping crate.

PAVN crew adjusting the range on a Soviet 82mm mortar. Emering collection.

Captured ChiCom Type 53 7.62mm LMG (copy of the Soviet DPM LMG). NVVA Museum, Chicago, IL, photo collection.

PAVN officer lectures air defense crew on antiaircraft operations. Note the water cooled ChiCom Maxim Type 24 HMG in an antiaircraft configuration (rear). The officer has a camouflage ring on top of his sun helmet and Soviet Bloc leather ammo pouches on his combat belt. Emering collection.

3

HEADGEAR

The PAVN and VC made use of an extremely wide range of headgear.

The early Viet Minh forces favored a very distinctive wooden helmet recognizable by its unusual shape. These helmets were made of woven bamboo and were worn with a variety of plastic covers with netting or camouflage added. They were issued to all ranks and worn with the early version of the red PAVN cap badge only by officers. It was also not uncommon to see the conical shaped straw ("coolie") hat being worn in the field. Sun helmets (referred to as pith helmets by Americans) were a rarity during this period.

As the armed forces matured, officers began wearing service dress peaked caps. Three distinctive cap badges, reflective of rank (General Officer, Field Grade Officer and all other ranks), and branch designators as denoted by color and/or device, i.e. wings on blue background for the Air Force and an anchor on a dark green background for the Coastal Defense Units (PAVN Navy) were worn at all levels starting in 1958. Sun helmets, resembling the European style pith helmets, were made of cardboard impregnated with resin, and came into vogue during this period, replacing the bamboo helmets of the Viet Minh. These sun helmets came in shades of dark green, tan, brown and in the case of the PAVN Navy, white. Copies of the sun helmet were actually made in Taiwan for use by Allied reconnaissance teams. The "originals" are distinguishable by their internal rubberized webbing system. There was also a flatter version of the sun helmet, usually covered with waterproof plastic, which was also in use at the same time.

During the Second Indochina War, it was common practice for North Vietnamese forces to remove their cap badges when crossing over into South Vietnam in order to maintain the guise that the PLAF was a separate fighting force. Cap badges were often found hidden in the shirt pockets or field packs of slain soldiers when searched by Allied forces.

In 1982, cap badges were standardized for all branches of the PAVN. The army's red cap badge became the new standard, with the exception of certain security police units, organized under the Ministry of the Interior. These forces continue to wear the distinctive Cong An cap badge.

In addition to the European Colonial style sun helmet (which was one of the few items actually manufactured in North Vietnam) and the formal peaked cap, a wide variety of soft covers and "boonie" hats were worn in the field, among these was a soft, floppy field hat reminiscent of the WW II Japanese army cap.

Original Viet Minh cap badge used prior to 1958.

Standard PAVN cap badge. This headgear insignia has been worn by all branches of PAVN and all ranks below field grade officer since 1982.

Distinctive Air Force cap badge used between 1965 and 1982.

Distinctive Coastal Defense (Navy) cap badge used between 1965 and 1982.

PAVN armored personnel favored the use of the distinctive black padded Russian tanker headgear.

Steel helmets, imported primarily from Poland, but also from Czechoslovakia, East Germany and Russia, were used primarily by self defense forces, antiaircraft and artillery crews and truck drivers. In 1982, a distinctive peaked service cap for General grade officers was introduced, which added a gold braid at the base of the cap's visor and a Soviet style, wide red band of cloth behind the distinctive General grade officers cap badge with its gold pine branches. After 1990, the standard PAVN cap badge was also affixed to the steel helmets. A wool-pile cap was issued for duty along the northern border with China and in the Central Highlands and was usually worn under some other form of headgear such as the sun helmet.

Viet Minh bamboo helmet with Viet Minh cap badge worn by officers only. Circa 1954.

PAVN tan sun "pith" helmet.

Viet Cong guerrilla fighter with plastic covered bamboo helmet. Note the wear of commercial shower sandals as foot gear. National Archives.

An Ninh security force (dark green) sun helmet.

Eastern European steel helmet with camouflage netting used primarily by antiaircraft crews and truck drivers.

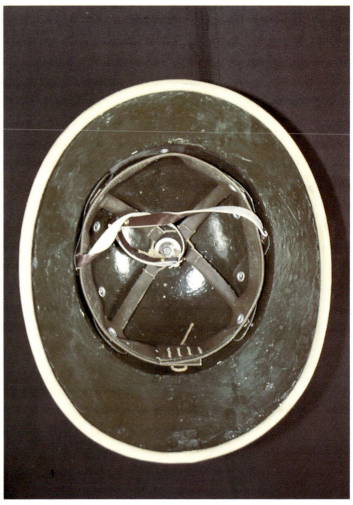

Original rubber webbing from a PAVN sun helmet.

Woven straw (17 inch) conical hat used in the field (but not in combat) by both PAVN and NLF forces.

Black NLF boonie hat.

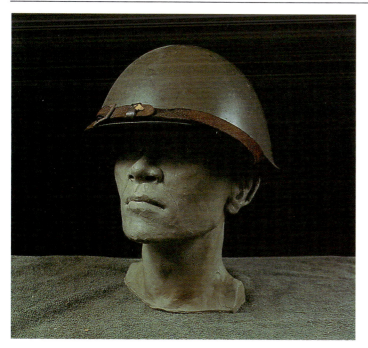

Eastern European steel helmet.

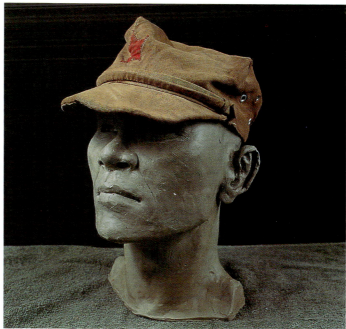

ChiCom advisor's field cap. China claimed that more than 300,000 of its forces served along side of the PAVN during the Second Indochina War. The PAVN later refuted this number, but not the fact that ChiCom forces served with them. Two ChiCom advisors were killed in the field along with seven members of a PAVN command element by MACV-SOG Reconnaissance Team (R/T) Maine in the spring of 1967 while they were attempting to hunt down the MACV team. Another ChiCom advisor was killed by R/T Washington on December 1, 1970.

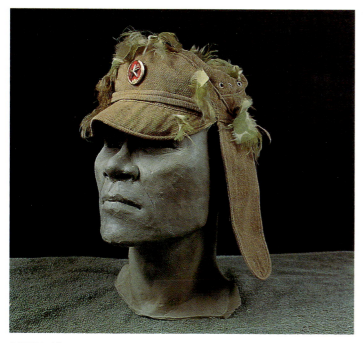

PAVN field cap variation with ear flaps reminiscent of the WW II Japanese army field hat. It was also worn by paramilitary and militia units.

Flatter version of the sun helmet with camouflage cover.

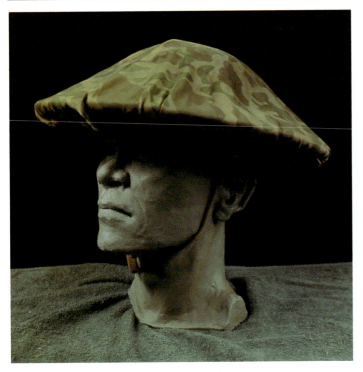

Another version of the flatter sun helmet with plastic camouflage cover.

Armor crew member's padded helmet.

NVA medic's steel helmet.

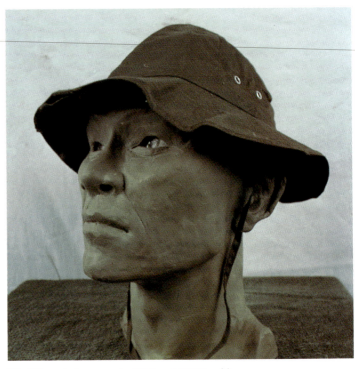

Khaki boonie hat worn by NLF and PAVN soldiers.

PAVN peaked cap now worn by all ranks below field grade officers on formal occasions. Peter Aitken collection.

Popular pile cap favored by units assigned to cold weather locations such as the ChiCom border region or the Central Highlands. Since 1982, the Cong An cap badge is worn only by select Ministry of Interior police units.

Rare PAVN tropical white cotton sun helmet with white chin strap.

General grade officer's formal peaked cap (reflective of the former Soviet Union style) with distinctive General grade cap badge. Emering collection.

PAVN Coastal Defense officer's peaked cap with unusual embroidered cap badge.

PAVN Coastal Defense peaked cap with distinctive cap badge used between 1965 and 1982.

Light green boonie hat.

Older wide brimmed tan boonie hat.

PAVN Coastal Defense "Donald Duck" cap, which reads: "Hai Quan Viet Nam" (Vietnamese Navy).

Tan boonie hat commonly worn by PAVN forces in the field.

Wide brim boonie, favored by both male and female combatants.

Gray boonie hat with rare, small red enamel star.

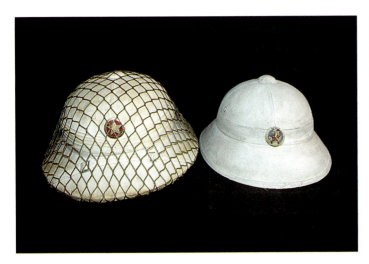

Rare tropical white cotton sun helmets. Helmet (left) has Viet Minh cap badge and camouflage netting cover. The helmet (right) has the PAVN Coastal Defense (Navy) distinctive cap badge, which was used between 1965 and 1982.

Camouflage ring used to attach branches and other camouflage materials to the sun helmet.

4

UNIFORMS AND INSIGNIA

Until supplies from China started moving into North Vietnam in bulk, uniforms, if any, were a purely haphazard affair. Prior to 1958, the operative term "catch as catch can" seemed to apply. When uniforms were worn, they had a decidedly Chinese character to them. Most distinctive to the period prior to 1958 was the bamboo style helmet with camouflage netting stretched over it and a noticeable lack of rank or branch designations. Only officers wore the PAVN helmet insignia, as previously mentioned. Subsequent to 1957, greater formality was introduced, but to the observer, uniform dress still appeared to be a haphazard affair.

Uniforms can also be divided into three periods: (1) prior to 1958; (2) 1958 through 1981, when an initial uniform code was adopted; and (3) subsequent to 1981, when a revised set of uniform regulations was adopted.

In 1958 PAVN began an initial modernization effort to standardize its uniforms. Khaki, dark green and tan field uniforms became prevalent, but wide varieties in color continued to prevail, especially during the Second Indochina War when materials were scarce. The standard issued field uniform (usually dark green) consisted of a long sleeve shirt with two button down breast pockets for enlisted men and four button down pockets for officers and NCOs. Matching trousers had three pockets and those worn by NCOs and enlisted ranks could be gathered at the ankle using a button and loop device. A web belt with a brass or aluminum buckle was worn with this uniform. Formal uniforms for general and field grade officers were authorized, but standards seemed lax and wide scale variations in color continued to prevail. During this period, the PAVN adopted a formal rank and branch designation structure. In addition to shoulder board rank designations for officers and non-commissioned officers (Senior General through Corporal), the PAVN adopted the following 1958 branch insignia, which was worn on collar tabs:

VC black pajama uniform with checkered scarf/towel popular among VC guerrilla units. Emering collection.

Branch	Device
Armor	Tank
Artillery	Crossed cannons
Cavalry	Crossed rifle and sword over horseshoe
Chemical	Radiation ray inside coil
Engineer	Crossed pick and shovel over half cogwheel
Medical	Encircled red cross
Military Justice	Crossed swords and shield
Ordnance	Crossed rifles over a cogwheel
Quartermaster	Encircled hoe with rice field
Signal	Encircled radio waves
Transportation	Steering wheel over truck suspension spring

Hero of the National Liberation Forces, Ta Thi Kieu, began her military career as a VC Squad leader in Bentre Province. She later rose to the role of international representative of the NLF. She is wearing a black pajama outfit with a checkered scarf/towel used as an improvised belt. She carries captured U.S. .30 caliber M-1 Garand rifles. Emering collection.

NLF main force squad on patrol armed with carbine rifles. Note the "mix and match" casualness of their uniforms and the ox drawn cart in the rear. Emering collection.

Infantry personnel wore plain collar tabs. In 1965, the Air Force instituted its own branch insignia metallic wings on a blue background. Coastal Defense (PAVN Navy) units also adopted a distinctive branch insignia a red anchor on a dark green background.

During this same period, the NLF forces favored the use of black, gray or brown pajama style field uniforms. These pajamas were common in South Vietnam and were procured locally by NLF forces. They came in both male and female versions. The long or short sleeve pajama shirts were collarless and the pants usually had an elasticized waist. VC units usually favored the use of the Ho Chi Minh sandal (as opposed to the PAVN rubber and canvas combat shoe). The base of the sandal was constructed from used tire treads and rubber or cloth strips were used to fasten the base to the foot. They also used the conical "coolie" hat, but usually not during combat action, favoring instead dark green or black "boonie" caps.

After 1970, the PLAF main force units, staffed primarily by PAVN soldiers, abandoned the civilian black pajama style uniform and attempted to standardize the use of black, brown or tan cotton uniforms (VC style pajamas). The shirts had collars and two button down breast pockets. Although no rank or branch designations were worn, officers could be identified by their shirts (usually worn with epaulets), their pistol belts and the number of pens in their shirt pocket. On rare occasions, NLF soldiers would be observed wearing a distinctive cap badge (sometimes pinned to their shirts), similar to that of the PAVN, but utilizing the NLF's blue and red color combination. PLAF guerrilla units still made due with whatever was available, including homemade captured U.S. and South Vietnamese clothing.

In 1982, the new field and dress uniform code attempted once and for all to standardize PAVN uniforms. Uniforms were hence-

PAVN squad on patrol circa 1970. Note the greater degree of formality to the uniforms, such as the ChiCom canvas and leather magazine chest pouches and the rubber and canvas combat shoe. The lead man carries a Czech ZB-26 LMG. Several members of the squad are wearing the PAVN cap badge on their sun helmets. Emering collection.

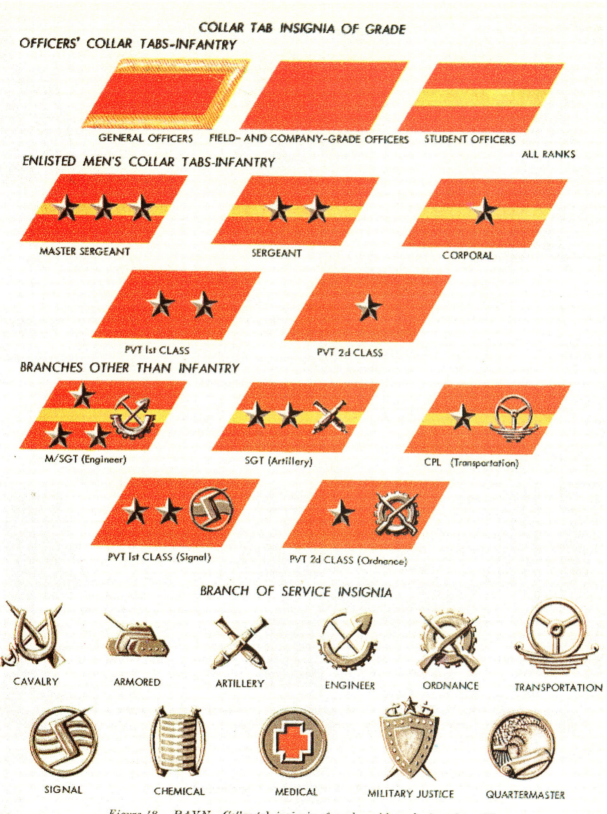

Figure 48. PAVN—Collar tab insignia of grade and branch of service. (U)

PAVN branch insignia in use prior to 1982. U.S. Army photo.

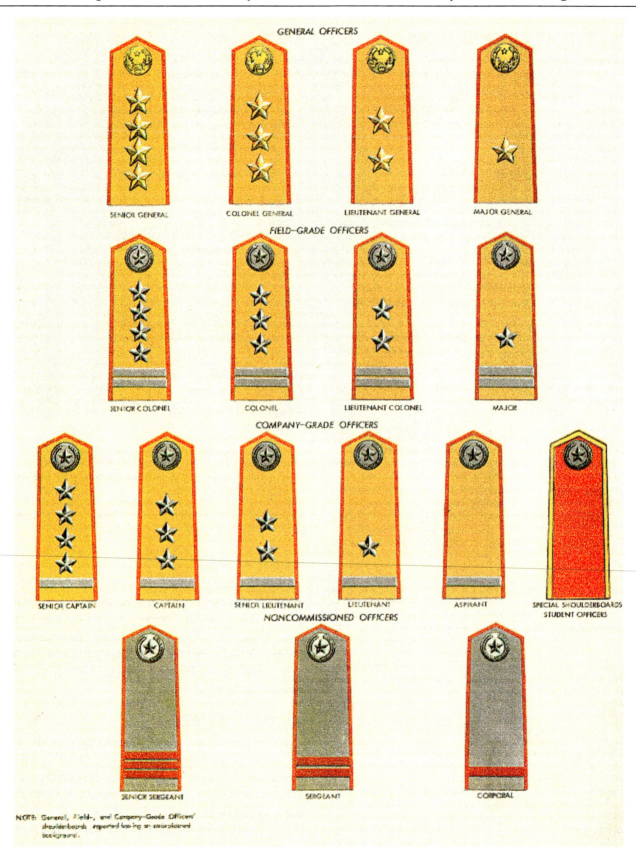

PAVN rank insignia in use prior to 1982. In 1982, rank insignia was extended to all levels and not just officers and NCOs. U.S. Army photo.

Well decorated (left to right) PAVN Artillery Officer with veteran Dac Cong (Sapper) Senior Captain, Signal Corp Captain and Transportation Major. The mixture of uniform jackets and peaked cap badges worn by these veterans reflects their casualness. The two officers in the center wear the distinctive field grade officer's cap badge, which was introduced in 1982. The most senior officer (right) wears the standard cap badge. The two officers (center) wear the closed neck four pocket tunic. The two officers (extreme right and left) are wearing the winter dress, single breasted tunic with open collar and notched lapels (note the shirt color and tie variations). Peter Aitken collection.

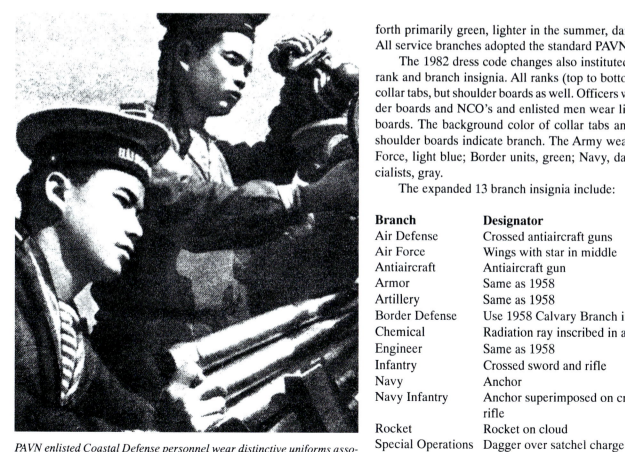

PAVN enlisted Coastal Defense personnel wear distinctive uniforms associated with the Navy Branch, including the "Donald Duck" cap, which is inscribed "Hai Quan Viet Nam" (Vietnam Navy) circa 1970. Emering collection.

forth primarily green, lighter in the summer, darker in the winter. All service branches adopted the standard PAVN red cap badge.

The 1982 dress code changes also instituted a new system of rank and branch insignia. All ranks (top to bottom) wore not only collar tabs, but shoulder boards as well. Officers wear yellow shoulder boards and NCO's and enlisted men wear light gray shoulder boards. The background color of collar tabs and the edge of the shoulder boards indicate branch. The Army wears scarlet; the Air Force, light blue; Border units, green; Navy, dark blue; and Specialists, gray.

The expanded 13 branch insignia include:

Branch	Designator
Air Defense	Crossed antiaircraft guns
Air Force	Wings with star in middle
Antiaircraft	Antiaircraft gun
Armor	Same as 1958
Artillery	Same as 1958
Border Defense	Use 1958 Calvary Branch insignia
Chemical	Radiation ray inscribed in a hexagon
Engineer	Same as 1958
Infantry	Crossed sword and rifle
Navy	Anchor
Navy Infantry	Anchor superimposed on crossed sword and rifle
Rocket	Rocket on cloud
Special Operations	Dagger over satchel charge

Airborne specialist insignia, adopted in 1982.

PAVN air force branch insignia, adopted in 1982.

Brown and tan leather dress uniform belts.

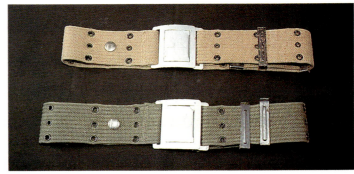

Tan and green combat belts.

Close-up of combat belt buckle star. Some commentators have reported these stars painted red, but this may be just a form of "dressing up" done by an eager seller.

Viet Cong tan cloth combat belt.

Rare brown leather, perforated garrison belt. It is most likely of Soviet Bloc origin.

Thin leather dress belt with seldom seen buckle.

In addition to the foregoing branch insignia, 12 specialist insignia were adopted in 1982. These include:

Specialty	Designator
Airborne	Winged parachute
Art Troupe	Music note with lute
Logistics	Crossed sword and rifle over rice stalk
Mechanized Infantry	Armored vehicle over crossed sword and rifle
Medical	Same as 1958
Military Band	Crossed trumpet and flute
Military Justice	Same as 1958
Physical Education	Bow and arrow
Radar	Radar antenna
Signal	Same as 1958
Technical Sector	Compass over hammer
Transportation	Same as 1958

Even with the 1982 revisions, uniform variations, especially in color, are still noted. Although General grade officers were authorized to wear a distinctive red band on their peaked cap, similar to that worn by the then Soviet General officers, they do not always do so. The PAVN Coastal Defense (Navy) branch and the Air Force seem to favor distinctive uniform colors.

The wearing of official SRV medals was also proscribed in the 1982 regulations. Hero Decorations and Orders are worn on the left breast. Longevity awards, communist party badges, and non-Hero Decorations are worn on the right breast.

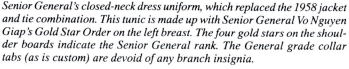

Senior General's closed-neck dress uniform, which replaced the 1958 jacket and tie combination. This tunic is made up with Senior General Vo Nguyen Giap's Gold Star Order on the left breast. The four gold stars on the shoulder boards indicate the Senior General rank. The General grade collar tabs (as is custom) are devoid of any branch insignia.

Armor crewman's green padded tunic and trousers. Note the four button down pockets. A Soviet tanker's helmet rests nearby.

Rare PAVN two pocket combat jacket with waistband.

PAVN enlisted two pocket field uniform shirt.

PAVN officer's four button pocket shirt with shoulder straps for securing the shoulder board rank insignia.

PAVN officer's four button pocket shirt without shoulder straps.

Gray medical corpsman's open neck, two pocket, notched lapel jacket and trousers (note the Vietnamese Red Cross badge on left lapel).

PAVN enlisted two button pocket combat shirt.

PAVN officer's four button pocket combat shirt.

PAVN air force officer's four button pocket combat shirt.

PAVN enlisted two button pocket uniform with sun helmet. The pants have a three button fly and hook closure at the waist band.

PAVN enlisted two button pocket combat shirt.

PAVN officer's four button pocket combat shirt and sun helmet.

PAVN officer's four button pocket uniform.

PAVN officer's four button pocket shirt.

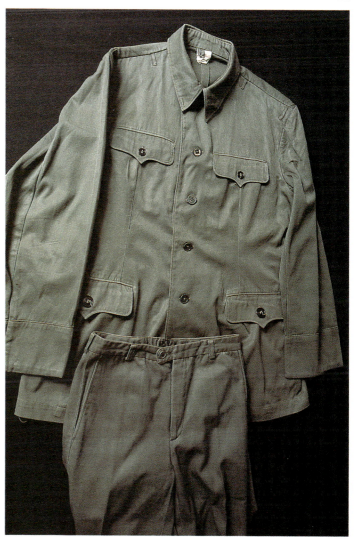

PAVN officer's four button pocket uniform. The three button pocket trousers have a zipper closure and button at the waist band.

Viet Cong guerrilla's pajama uniform.

Guerrilla black and white checkered scarf/towel.

PAVN Coastal Defense Petty Officer Third Class uniform.

PAVN Coastal Defense enlisted uniform (white shirt, blue trousers).

PAVN Coastal Defense enlisted blue uniform shirt and cap.

PAVN pullover T-shirt.

PAVN enlisted field uniform with two button down pocket shirt.

PAVN enlisted uniform with two button pocket shirt.

Close-up of button closure on cuff of PAVN enlisted uniform trousers.

Krama, a traditional Cambodian red and white checkered scarf/towel.

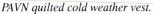

PAVN quilted cold weather vest.

PAVN air force NCO's utility shirt with 1982 branch designator on left breast.

PAVN in Space: PAVN Air Force Lieutenant Colonel Pham Tuan (left) with his Soviet counterpart just prior to their July 23, 1981, historic Soyuz 37 space flight, launched from the Baykonur City Cosmodome. Soyuz 37 linked up with the Soviet Space Lab, Salyut Six during its flight. Tuan had been credited with the shoot down of a U.S. B-52 during the Second Indochina War. Tuan is shown wearing a custom made Soviet space suit for the flight. Photo courtesy Gerry Schooler.

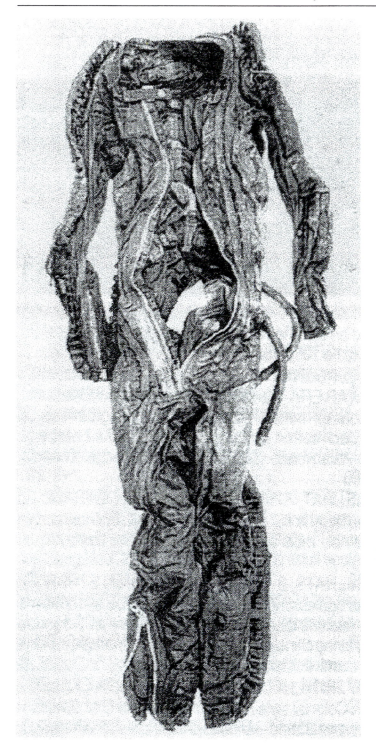

ChiCom manufactured green synthetic antigravity suit with pressure hoses and fittings. Front, arms and legs have zipper closures. The suit was used by PAVN MiG pilots during the Second Indochina War.

PAVN service ribbons (Rows are aligned from bottom left corner of image: Top row - first class, second row - second class and third and fourth rows - third or sole class). Institute of Orders, Hanoi.

5

FOOT WEAR

The PAVN and NLF were noted for their practical and comfortable foot wear. The PAVN forces favored the Bata boot, a rubber and canvas combat shoe (both high and low cut), which was adopted from the French Colonial forces. This shoe has proved so popular that eventually Vietnam found itself exporting the shoe to Eastern European Communist Bloc armies in the 1990s.

At the same time, the NLF and PAVN units operating in the warmer southern climates developed the Ho Chi Minh sandal, which reportedly used old rubber tires for soles and cloth or rubber strips for fasteners. When not available, imported rubber or plastic shower shoes were used. During the Second Indochina War, it would not have been unusual to encounter bare foot troops, especially among the VC units.

Officers wear both high and low cut leather shoes with their service dress uniforms. Armor crews have adopted the use of a low cut leather boot.

PAVN rubber and canvas combat shoe.

Ho Chi Minh sandals.

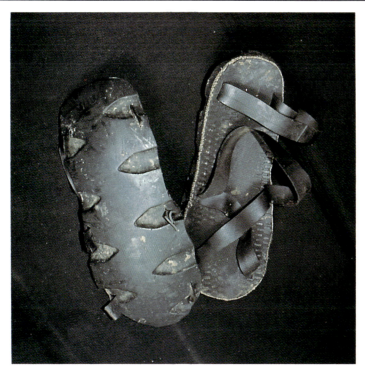

Ho Chi Minh sandals (variation).

Ho Chi Minh sandals (variation).

PAVN high top leather boots.

Black canvas combat shoe.

6

RADIOS

The early local guerrilla forces often operated without the benefit of electronic communications, but by the early 1960s with the infiltration of PAVN forces and the formation of main force NLF units, electronic means of communications for command and control purposes were an absolute necessity. The prophetic battle of Ap Bac on January 2, 1963 revolved around efforts of the South Vietnamese Army to silence a VC radio station, operating only 30 miles outside of Saigon. The PAVN and NLF turned primarily to their Soviet and Chinese allies as a source of field radios.

With the entry of the U.S. into the Vietnam War and the arrival of elements of the Army Security Agency and Naval Security Group, PAVN and VC radio transmissions became easy targets of intercept, decryption and translation. So successful were these two top secret U.S. military agencies, that in many cases PAVN field leaders were forced to utilize field telephone land lines to mask their positions and their intentions. The "point to point" wire communication links proved more difficult, but not impossible to monitor.

Throughout the entire Vietnam War, the most reliable means of enemy communication remained the personal messenger. During the final stages of the Second Indochina War, the Politburo sent its highest priority communications to their field leader, Senior General Van Tien Dung, via personal courier down the rugged and dangerous Ho Chi Minh Trail.

During the Second Indochina War, VC and main force guerrilla units had limited access to short wave radio equipment. This limited equipment, consisting primarily of Soviet and ChiCom gear, was augmented by field telephone wire systems, often depending on captured or abandoned U.S. wire to establish communication

Signal Corps Branch insignia.

Soviet R-105 field radio and antenna.

Main force VC soldier with "water lily" style bush hat using captured U.S. PRC 10 field radio.

ChiCom copy (Type 59) of the Soviet R-105 infantry company backpack field radio.

ChiCom radio transceiver Type 71-B battalion single side radio with integrated back pack.

Soviet YM-2 radio.

ChiCom field telephone Model E0743 (with instructions in Chinese) in black Bakelite case.

ChiCom Type 50 radio dry batteries and carrying pouch.

Radio gear pouches. The tall pouch (left) holds the Soviet Type 292 extended antenna kit.

links. The use of these systems, oddly enough, often resulted in fires during the dry season caused by shorting wires.

Among the primary field radios used by the PAVN and main force NLF units were the Soviet R-105, and the ChiCom Type 71B, which was later upgraded to the ChiCom Type 63. These radios could service both voice and continuous wave (CW). The Type 71-B operated on 1.9 to 7.2 megacycles and had an effective range of 30 miles. The Type 63, which was similar to the U.S. PRC 25, operated from 1.5 to 6 megacycles with an effective maximum range of 30 miles. For longer range short wave communications, the ChiCom 102E was utilized, effectively doubling the range of the other models. All radios used by the enemy had integrated carrying harnesses.

The enemy also made use of captured U.S. PRC 10, 25 and 77 radios, although these were not compatible with Soviet and ChiCom radio equipment and saw limited use by the enemy.

Field telephone sets utilized consisted of U.S. TA-312s, Soviet TA 1-43s and ChiCom E0574s (E0743) and Q-07.1s. The E0574 was the export model and the E0743 the domestic model. The only difference is that the E0574 had instructions printed in English and the E0743 in Chinese. Both operated on dry batteries powered by an integral hand cranked generator and had an effective range of 15 to 20 miles.

The enemy, particularly the NLF, also made wide scale use of much less sophisticated means of signaling such as the use of bugles, drums, whistles and small arms fire to sound the alert.

Field telephone line and pouch.

ChiCom manually operated field radio generator with 1962 manufacture date.

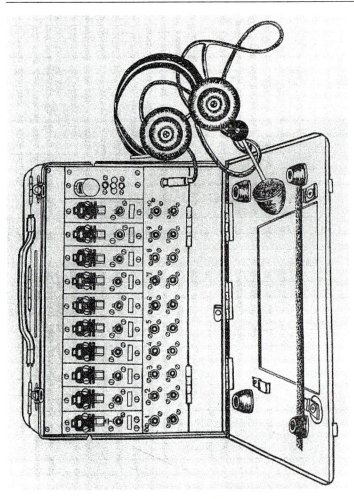

ChiCom ten line field switchboard. The board uses keys in lieu of connecting cords. It weighs approximately 20 pounds and can accommodate eight users simultaneously. Chien Cu Part II.

Main force NLF soldier sounds the alarm with a Czech made bugle. Note the unit banner hanging from the bugle. Goodrow collection.

Captured VC hollow signaling drum.

7

CANTEENS

The practical North Vietnamese made use of just about every piece of equipment they could scavenge. The canteen, which was an absolute necessity in the field, was no exception. Captured canteens from the Japanese, French, Americans and South Vietnamese were used. In addition, canteens were supplied from Russia, China and other Soviet Bloc countries. Canteens were made from materials ranging from aluminum to plastic to rubber.

Early canteens were usually made of one liter black or dark green aluminum bodies. These canteens had black or dark brown Bakelite (a form of plastic resin) threaded caps. Another early style of canteen came with a cotton cover to help reduce trail noise. These canteens gave way to green molded plastic canteens manufactured in China, which are readily identified by the green star molded into the body and the brown Bakelite threaded cap.

The PAVN eventually standardized the use of an olive painted, aluminum canteen with a brown Bakelite cap and a skeleton style web holder, reminiscent of the British forces. In many cases, the olive paint has been chipped away leaving only the metallic finish on the canteen.

ChiCom made "standard" NVA aluminum canteen with black Bakelite cover in skeleton web carrier.

ChiCom standard aluminum canteen version with olive drab paint.

Green plastic canteen and cloth pouch. Canteen has a Bakelite cover.

Thin aluminum canteen in WW II Japanese canteen pouch with lace up sides. The top has a metal stopper.

Green molded plastic ChiCom canteen with raised five pointed star within a circle on side and plastic cap. This canteen became a PAVN standard issue item during the later stages of the Second Indochina War.

Close-up detail of the raised green star.

Standard style aluminum canteen variation with an aluminum threaded top in web carrier.

Olive drab aluminum canteen with light plastic stopper in khaki quilted cover.

Padded canteen carrier with black Bakelite top. The padded cover has a belt loop sewn on the reverse.

Olive drab plastic canteen in web carrier with black Bakelite threaded top.

Reddish brown plastic canteen with flat plastic cover in web carrier.

Squat pumpkin shaped plastic canteen with belt clip fastener.

Aluminum bottle shaped canteen with red plastic top in plastic web carrier.

Standard style aluminum canteen variation with slotted aluminum top for web carrier to pass through.

Aluminum canteen with slotted top variation in crudely fashioned web style rope carrier.

Olive drab aluminum canteen with aluminum slotted top in web carrier.

Double size olive drab plastic molded canteen with black Bakelite cap in unique web style carrier. This canteen may have been used by field medics.

8

MESS KITS

Mess kits and utensils were lightweight and practical. Nearly every cadre carried some form of tin eating utensil (usually a deep spoon), a small cooking tin for heating water and a rice bowl. The PAVN soldier could subsist on a tube of dry food (primarily rice) for up to seven days. The rice would be cooked over an open flame using the small cooking tin and boiled water. Cooked rice was made more appealing by the addition of nouc mam (fermented fish sauce).

Aluminum cooking tin, spoon and porcelain glaze (center) and aluminum rice bowls.

Flat style aluminum mess tin.

"Lunch pail" style, 3 piece aluminum mess kit with carrying handle.

Light green "lunch pail" style, 3 piece (variation) aluminum mess kit with wire carrying handle.

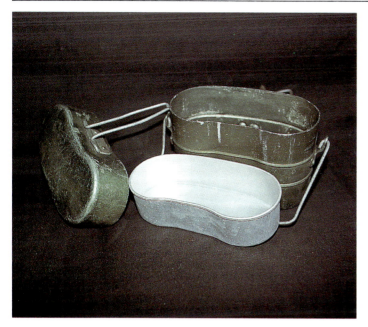

The individual components of the 3 piece "lunch pail."

Three piece mess kit with canvas carrying strap.

Green porcelain rice bowl with small metal ring for attachment to the combat gear.

White porcelain cup/bowl with Japanese style fish drawing.

Rice rations being prepared from a common cooking pot in jungle camp. F. C. Brown Collection, NVVA Museum, Chicago, Illinois.

9

MEDICAL KITS

Medical care for NLF and PAVN units operating in South Vietnam ranged from meager to nonexistent. There were no helicopter "dustoffs" available and the wounded were transported by rudimentary means ranging from stretcher bearers to ox carts. In some cases, they were literally drug off the battle field. Field hospitals were located underground in caves or tunnels or deep inside the jungle and were staffed by often poorly trained personnel, operating with only bare essentials. Conditions were often extremely primitive with parachute nylon being used to cordon off otherwise unsterile operating areas. In many ways, the enemy's practice of field medicine compared to the type of medicine practiced during the U.S. Civil War, with the ultimate solution being amputation of wounded limbs. Although medical personnel were at a premium, each company did have at a minimum one medical corpsman trained in first aid. It was, however reported that the medical treatment received was perceived as being of high quality by the NLF and PAVN soldiers.

One outstanding exception was Viet Cong physician and Hero of the Liberation Armed Forces, Colonel Doctor Vo Hoang Le. Dr. Vo was head of the National Liberation Front's Medical Section IV, located in the tunnels at Cu Chi. He was said to have worked miracles under the most primitive conditions. Assisted by his nurse-wife, Nguyen Thi Tham, he was even purported to have performed delicate brain surgery using an ordinary household drill. He also made wide use of natural substances such as honey, which he used as an antiseptic. He and his wife were both highly decorated for their efforts.

Wounds and diseases, such as malaria, yellow fever, beriberi, tuberculosis, dysentery and hepatitis received expedient treatment using whatever was available. Often medicine and vitamins, used to treat nutritional deficiencies, were in extremely short supply, particularly among those units operating in the South. North Vietnam relied heavily on Soviet and ChiCom aid, as well as assistance from Cambodia and Laos and upon humanitarian aid from otherwise neutral nations, such as the Scandinavian countries and Switzerland, and humanitarian organizations to bolster its meager, compared to the size of its army, medical infrastructure. In extreme cases, enemy units in the South depended on U.S. drugs and medicines, procured on the black market in Saigon. Items such as sanitary napkins proved to be an acceptable alternative for use as field dressings.

Captured field gear includes ChiCom first aid kits, field bandage packs, legend drugs such as morphine, which were packed in a small heat sealed glass viles, and rudimentary medical instruments. Almost every captured medical kit contained vitamins for treatment of dietary deficiencies and malaria.

NVA Corpsman's (Quan Y) badge.

ChiCom first aid kit.

Improvised pouch with field bandages.

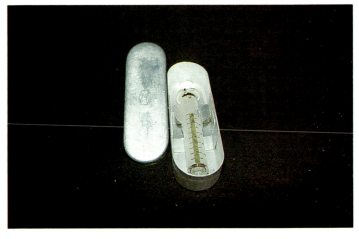

Hypodermic needle in aluminum carrying case.

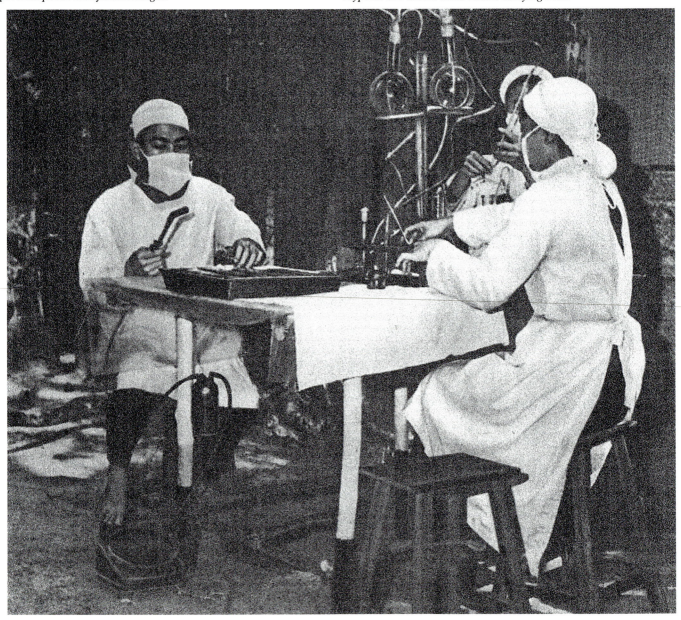

Crude NLF field hospital. Emering collection.

Hero of the Liberation Armed Forces Colonel Vo Hoang Le, Chief of the NLF's Medical Section IV at Cu Chi. F. C. Brown Collection, of the NVVA Museum, Chicago, Illinois.

Soviet bandage pack.

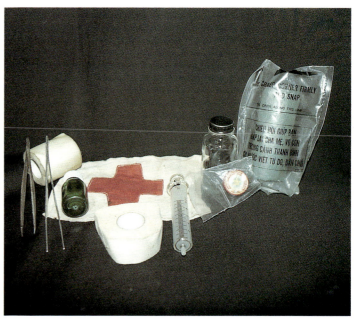

Viet Cong medical instruments, including hypodermic needle, bandages and medicines.

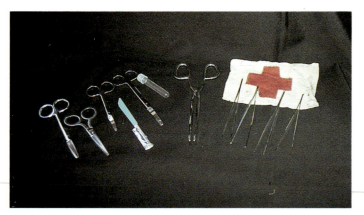

Viet Cong medical instruments, including scalpel with turquoise foil wrapper and field medic's arm band.

Enemy single dose, heat sealed glass medicine viles.

Medical facility badges: (1) BS - doctor; and (2) YT - nurse.

Soviet WW II field medical pack.

ChiCom field medical pack.

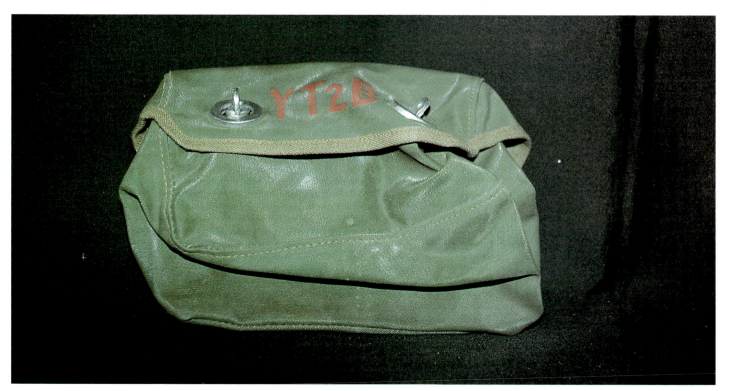

ChiCom field aid pack.

10

GAS MASKS

Just how widespread the use of chemical warfare (CW) was in Vietnam is difficult to determine. Certainly the allies made use of various types of riot control grenades, including CN (a "tear" gas), DM (a vomiting agent) and CS (a more powerful form of "tear" gas). Allied forces even pumped (as Radio Hanoi claimed) poison gas into the NLF tunnel complexes at Cu Chi in the South (actually acetylene gas was pumped into to the tunnels with leaf blowers). In fact the Allied chemical warfare effort seems to have been centered around the areas of Cu Chi and the Iron Triangle. It should be noted that the enemy viewed the defoliant known as Agent Orange as a poison gas. Whatever the reason, enemy soldiers, especially PAVN fillers, carried gas masks as part of their equipment.

The more sophisticated canister type gas masks (if not captured American gear) seem to be primarily of Soviet or ChiCom origin. The ChiCom two piece Type 66 and the Soviet ShK-1 (or the smaller version ShM-1 favored by the diminutive Viet Cong) were the most sophisticated masks in use.

The most common protection against CW were locally produced, simple, nylon "bag" masks, which were often "manufactured" from parachute material. These bag masks contained only a simple filter made from cotton, thin charcoal pellets and linen. In many cases, a slice of lemon was added to the filter. These elasticized masks could be hastily pulled over the head in case of attack. While not as sophisticated as the canister type mask, their advantages were that they were light and could be easily stored for carrying. When masks were not available, the enemy resorted to placing urine soaked cloths over their nose and mouth.

It is also known that the enemy possessed gas grenades, but how extensive their use was, if any, remains undocumented. Primary reference sources relating to the issue of the use of chemical warfare during the Second Indochina War are enemy propaganda publications, produced in Moscow.

The bottom half of the two piece ChiCom Type 66 gas mask. The bottom half is composed of a molded, green rubber nose and mouth cover connected to a green metal purification canister. Missing is a pair of molded, green rubber eye goggles with plastic lenses.

One piece green plastic bag mask. The inside has a molded plastic nose and mouth cover. The basic filter system is composed of gauze and charcoal.

One piece brown plastic bag mask. The mask uses a gauze and charcoal filter. There are no molded parts.

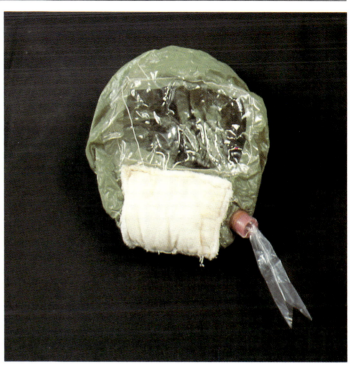

Extremely simple green plastic bag mask without molded internal nose and mouth cover.

Soviet ShK-1 gas mask with concertina type canister and carrying pouch. The mask is elasticized for a snug fit.

PAVN gas mask with filter on left side. Note the star design on the outside of the filter.

11

GUN ACCESSORIES

One common gun accessory was the universal holster, the ChiCom Type 54. It was made of reddish brown leather and contained an integrated magazine pouch as well as a compartment for its cleaning rod. Additional magazine pouches were usually affixed directly to the combat belt itself.

All other combatants carried cleaning tools specific to the weapons assigned to them. These tools usually consisted of a 3 inch to 4 inch cylinder, a bore brush, a patch ram and a cleaning rod. In addition, rifle oil and cleaning solvent were carried in "oilers." These were dual spout, dual compartment tin containers, approximately 2 inches high. The tins were marked with Chinese or Cyrillic acronyms: "Alkaline Solvent and Gun Oil." The shapes of the oilers varied widely from rounded bottoms to flat bottoms. Several different examples are shown, including larger flask like containers of gun oil used for heavier weapons such as machine guns.

Other common acquisitions include cartridge pouches and the cartridge packs themselves, usually well marked in Chinese.

Various gun cleaning supplies and brown Warsaw Pact expandable leather cartridge packs with brass hardware (center). Pack on left contains cleaning swabs. Various oilers in the rear hold cleaning fluid and gun oil, including the large olive drab metal can. The small glass container (background) is a trail lantern.

ChiCom Type 54 "standard" leather holster with dual leather clip pouches.

Round metal tin of wax and ladle used for sealing fuses in grenades and mines (left); and large olive drab metal can of gun oil for heavy weapons (right).

Close-up of dual compartment ChiCom round bottom "oiler" which reads in Chinese: "alkaline solvent" and "gun oil."

Close-up of dual compartment Soviet round bottom "oiler" with the Cyrillic letters for "alkaline solvent" and "gun oil."

Various inner packs of ChiCom cartridges in brown paper and cardboard wrappers. The markings indicate caliber, cartridge case, type and quantity of the ammunition.

ChiCom pistol cleaning rod from Type 54 holster.

ChiCom RPG optic and carrying case with shoulder strap.

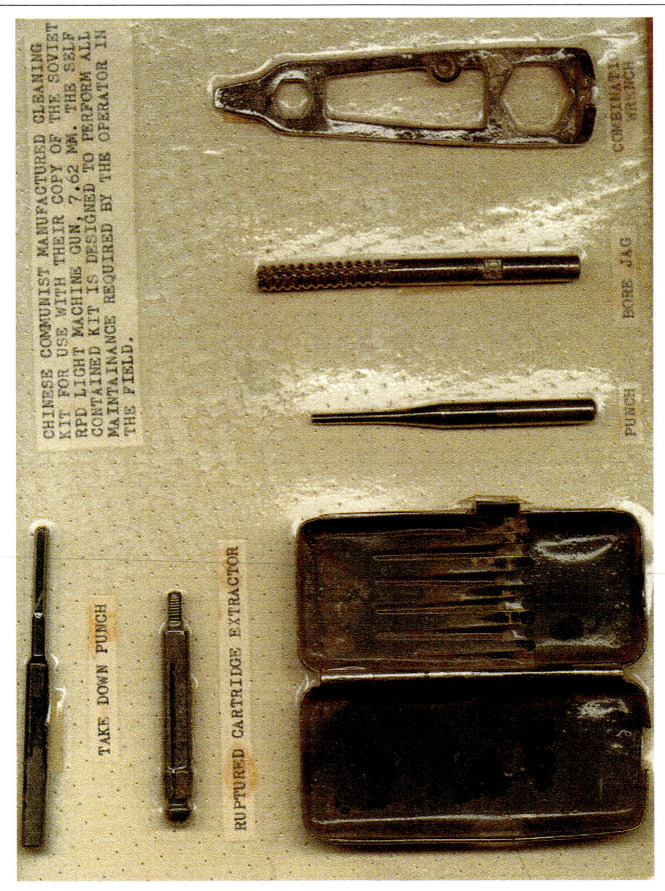

CHINESE COMMUNIST MANUFACTURED CLEANING KIT FOR USE WITH THEIR COPY OF THE SOVIET RPD LIGHT MACHINE GUN, 7.62 MM. THE SELF CONTAINED KIT IS DESIGNED TO PERFORM ALL MAINTAINANCE REQUIRED BY THE OPERATOR IN THE FIELD.

COMBINATION WRENCH

BORE JAG

PUNCH

TAKE DOWN PUNCH

RUPTURED CARTRIDGE EXTRACTOR

ChiCom cleaning kit for the Type 56 (RPD copy) LMG.

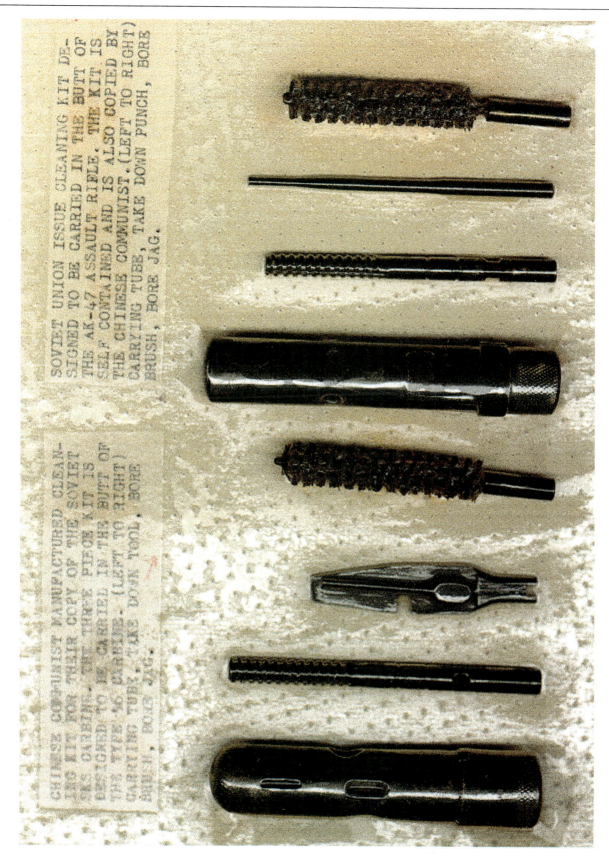

SOVIET UNION ISSUE CLEANING KIT DE-SIGNED TO BE CARRIED IN THE BUTT OF THE AK-47 ASSAULT RIFLE. THE KIT IS SELF CONTAINED AND IS ALSO COPIED BY THE CHINESE COMMUNIST. (LEFT TO RIGHT) CARRYING TUBE, TAKE DOWN PUNCH, BORE BRUSH, BORE JAG.

CHINESE COMMUNIST MANUFACTURED CLEAN-ING KIT FOR THEIR COPY OF THE SOVIET SKS CARBINE. THIS THREE PIECE KIT IS DESIGNED TO BE CARRIED IN THE BUTT OF THE TYPE 56 CARBINE. (LEFT TO RIGHT) CARRYING TUBE, TAKE DOWN TOOL, BORE BRUSH, BORE JAG.

ChiCom cleaning kits for the Type 56 (SKS copy) and Type 56 (AK-47 copy) assault rifles.

Additional view of the optic case (left) and "jungle made" magazine pouch.

Jungle made magazine pouch and ammunition clips.

RPG-2 (B-40) chest pouch.

Close-up of the markings on the inside of the RPG-2 (B-40) chest pouch.

Rifle grenade pouch.

ChiCom Type 56 (AK-47) assault rifle chest pouch.

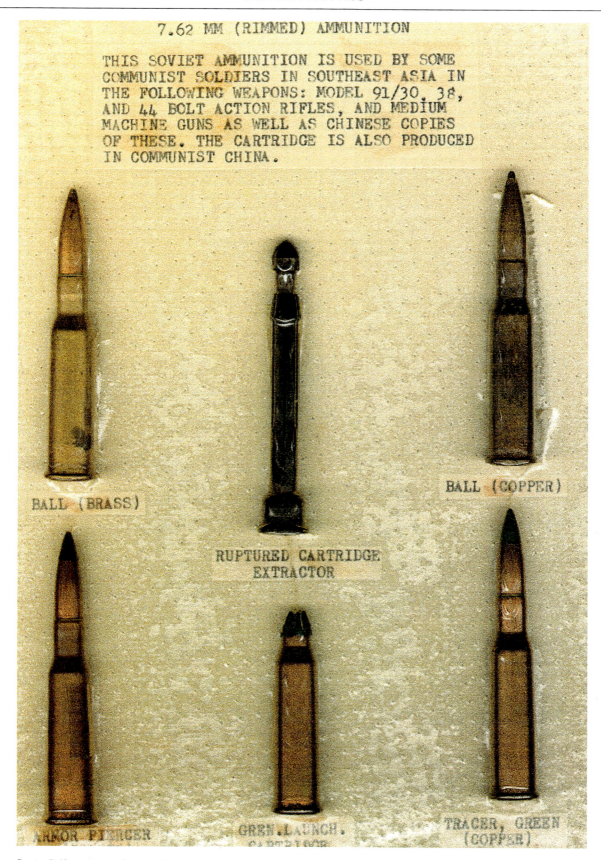

Soviet 7.62mm rimmed ammunition.

7.62 MM (M1943) AMMUNITION

CHINESE COMMUNIST AND SOVIET UNION AMMUNITION FOR
USE IN THE MORE COMMON WEAPONS CURRENTLY BEING USED
BY VIET CONG AND NORTH VIETNAMESE REGULARS IN SOUTH
EAST ASIA. THIS ROUND IS USED IN THE AK47 ASSAULT
RIFLE SERIES, AKM LIGHTWEIGHT ASSAULT RIFLE, THE SKS
CARBINE, AND THE RPD LIGHT MACHINE GUN AND THE CHIN-
ESE COMMUNIST COPIES OF THESE SOVIET WEAPONS.

BALL (BRASS)   BALL (COPPER)   GREN.LAUNCH.
                                   CARTRIDGE

ARMOR PIERCING   ARM.PIERCING TRACER   TRACER (GREEN)
                        (RED)

Soviet/ChiCom 7.62mm (M1943) ammunition.

Close-up of 1962 manufacture date on inside of the AK-47 chest pouch. This date is used to authenticate the chest pouch as original war era militaria.

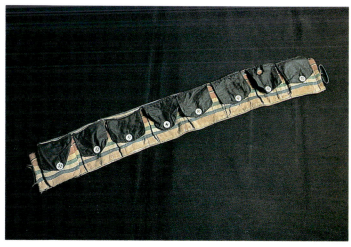

An extremely rare Montagnard made SKS (ChiCom Type 56) assault rifle chest pouch.

Leather and canvas ChiCom SKS pouch with waist strap.

PPSh 41 drum magazine canvas pouch with shoulder strap.

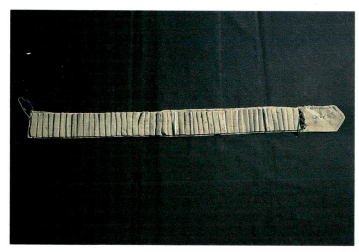

Unusual canvas "Pancho Villa" style bandoleer.

Four pocket canvas magazine pouch.

Canvas drum magazine pouch with leather strap.

Rare leather magazine pouch for the MAT-49 with shoulder strap.

Canvas stick magazine pouch with shoulder strap.

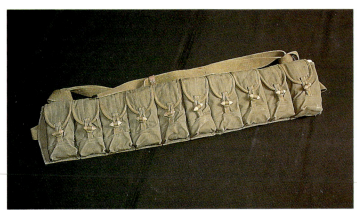

Ten pocket canvas magazine pouch with waist strap.

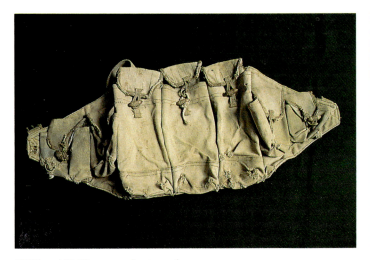

ChiCom AK-47 canvas chest pouch.

ChiCom 82mm mortar optical sight container with shoulder strap.

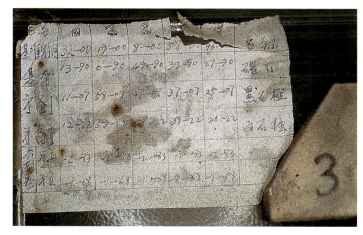

Elevation settings shown on inside of cover of the ChiCom 82mm optical sight container.

Four pocket canvas and leather chest pouch.

Rare MAT-49 leather stick magazine pouch with canvas carrying strap.

Five pocket leather magazine pouch with waist strap.

Jungle made .30 caliber magazine pouch with shoulder strap.

Rare RPG-2 rocket back pack with cleaning tools in the bottom of the pack.

Rare metal 60mm mortar bomb valise.

PAVN AK-47 magazine shoulder pouch with magazine and unmarked oiler.

Soviet LPO-50 flamethrower maintenance kit and pouch.

Khaki three pocket stick magazine shoulder pouch.

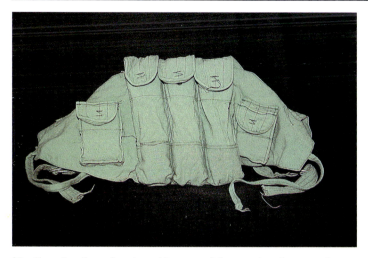

Viet Cong jungle made submachine gun stick magazine chest pouch.

ChiCom angle setting tool used for positioning the 122mm single shot rocket launcher.

Carrying pack for the warhead and fuse used with the 122mm rocket.

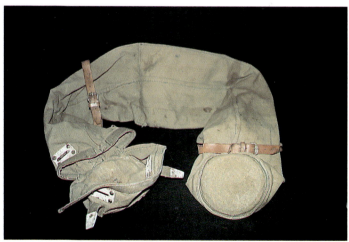

Carrying pack for the 122mm rocket body.

12

PERSONAL ACCESSORIES, KNIVES, AND TOOLS

As a matter of course, all enemy killed in action had their uniforms and field packs searched for intelligence purposes. As a result, a great deal of personal gear has made its way back to the collectors market. Whether taken by victorious troops as battlefield souvenirs or simply stockpiled in the rear, an amazing array of documents and gear, including field packs, rice tubes, belts, lighters, whistles, hand tools, and homemade and manufactured knives were collected. In at least one case, even an unknown, enemy soldier's "trail" art has made it back to the States. In many cases, such personal gear was "homemade" or improvised in crude jungle workshops. The variety of such gear reflects a high degree of ingenuity. Examples of such effort include a knife fashioned from the plexiglass windshield of a downed U.S. helicopter and a pack with carrying straps made from a U.S. Aid food bag. In other cases, many items of personal gear were donated by the Soviets and Chinese.

The manufacture of knives, which were carried by most enemy soldiers as tools more so than weapons, must have been one of the enemy's leading cottage industries. Most field knives were thin bladed, kitchen type instruments with wooden or bamboo handles and sheaths made from canvas or bamboo. An interesting knife type device had an eight inch curved blade at the end of an approximately 18 inch wooden handle. This tool was originally used by rubber plantation workers to score the rubber trees for sap collection. Many such knives were captured from VC units. The enemy also made use of short blade machetes for clearing jungle and/or cutting firewood.

In addition, a wide variety of identity papers and awards documents have also been collected and are routinely traded. Of most value to militaria collectors are such documents relating to the period 1960 through 1975. Many such documents were prepared on very fragile rice paper and in yet other cases on very ornate and elaborately decorated forms with official rubber stamped seals affixed. The stamps would be carried by senior officers in small water proof tins. In most cases, the captured documents are extremely fragile due to their exposure to harsh jungle conditions.

Standard PAVN field gear it would include a ChiCom rucksack, based on the French Colonial force's design. These rucksacks also originated from North Korea and Hong Kong. The rucksack would have a large draw string compartment and three smaller external button down pockets. Inside the rucksack would be at least one clean/dry uniform, a pair of black pajamas, two pair of underwear, an extra pair of sandals or rubber/canvas combat boots, a small packet of medicine, most likely antimalarial pills, and field dressings (if not carried in an external first aid kit). In addition, attached to a web belt might be such gear as a canteen, a first aid kit, rice bowl, an entrenching tool fastened by old pieces of inner tube, a knife or machete, possibly of the homemade variety and a two or four pocket ChiCom grenade pouch with mix and match grenades, depending on availability. In addition, most PAVN soldiers and some main line NLF troops, carried a chest pouch, which fastened like an apron with a neck loop and behind the back ties. These chest pouches were specific to various types of weapons and would include large pockets for additional magazines or rockets

Four "homemade" VC knives. Unsheathed knife (center foreground) has a bamboo scabbard.

VC machete on combat belt, ChiCom first aid kit (left), leather cartridge packs with belt loops and small knife (foreground) made from the plexiglass windscreen of a downed helicopter.

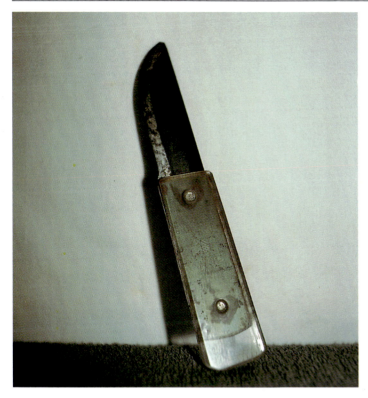

Close-up of the plexiglass knife.

Homemade pick with bamboo handle and forged iron head used as an entrenching tool.

and smaller pockets for additional cartridge boxes, gun tools and cleaning and oiling equipment. Shoulder bags and rice tubes completed the PAVN soldier's personal gear. NLF units would also utilize variations on this theme, but in many cases were much more lightly equipped.

Standard PAVN/VC entrenching tools with wooden handles and steel blades. These tools closely resemble those issued to ChiCom forces. Empty rice tubes used to carry dry food stuff rest at the base of the shovels.

Rubber plantation tool with 18 inch bamboo handle and forged iron blade. This tool was confiscated from rubber plantation workers by the VC and used to clear firing lanes.

NLF rubber stamps and tin holder. These stamps were used for official documents.

Entrenching tool with cord used to attach it to the combat belt. Note bullet hole in the blade.

Close-up of the rubber stamps.

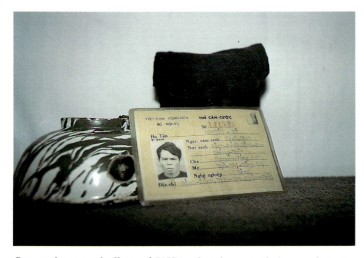

Captured personal effects of PAVN enlisted man include porcelain rice bowl and identity card issued by the Ministry of Interior Police. This soldier, Ty Nguyen, was a farmer before being conscripted.

Example of a VC medal, "Hero of the Assault" and corresponding rice paper award document often found among the personal gear of slain or captured enemy soldiers.

Standard PAVN rucksack with large interior and three exterior compartments. This rucksack has a rigid frame and leather straps.

NLF tan formless three exterior pocket rucksack with fabric ties. An unusual pair of motorcycles goggles are at the base of the sack.

Improvised woven straw basket backpack.

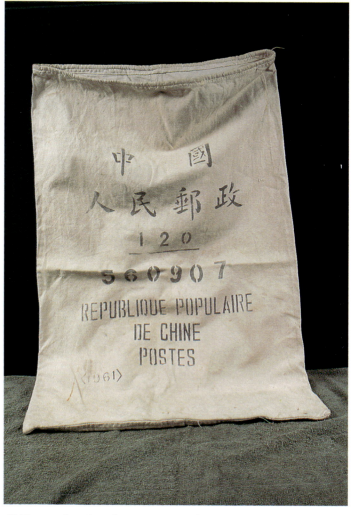

ChiCom mail sack used as an improvised rucksack.

Improvised blue rucksack with green two pocket stick grenade holder and an improvised woven basket grenade holder.

Standard rice tube, which held a week's dry rations. Soviet 82mm plastic propelling fuse (left) and 120mm HEAT fuse (right) are in foreground.

VC combat belt with (left to right) small field pouch, standard issue combat knife, aluminum canteen in padded pouch and dual grenade pouch with two ChiCom long handle stick grenades.

Combat belt with small improvised pouch foreground and (left to right in background) a small fabricated pouch, a rolled and tied piece of polyethylene used as a hammock or ground cover and an unpainted aluminum canteen with cloth cover.

Combat belt with (left to right) a fabricated knife and sheath, tied ground cover, a single ChiCom short handle grenade, fastened directly to the belt without a pouch, and an aluminum canteen with Bakelite top and padded cover.

Cigarette lighters and metallic whistle used as a signaling device.

Back (left to right) 12.7mm HMG cartridge casing, small pouch, 140mm rocket fuse (also adapted to the ChiCom 107mm rocket), small pouch and small tin case of wax used to seal grenade and/or mine fuses. Front (left to right) hacksaw used for grenade and booby trap making, two steel darts and pliers from a jungle workshop.

Two pocket grenade pouch (rear view) with belt loops for attachment to combat belt.

Four pocket grenade pouch with shoulder strap.

Improvised pack made from two 82mm ordnance cases lashed together with a U.S. Aid sack used for straps.

Improvised pack/storage case using a 60mm mortar bomb shipping container.

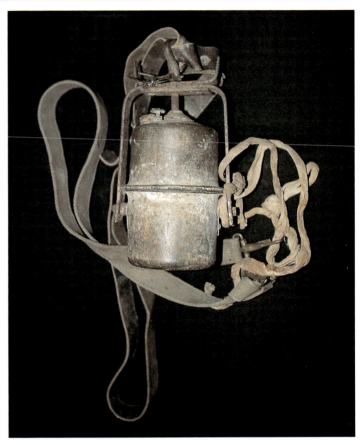

Carbide lantern used by Viet Cong tunnel dwellers.

Three crudely made leather grenade pouches with loops for attaching to the combat belt.

Trail hammock with hanging rope.

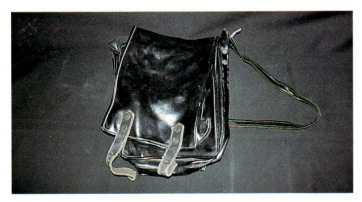

Rare black leather PAVN map case.

Two brown leather field pouches.

Examples of captured enemy documents (top to bottom): (1) NLF Contribution Receipt; (2) North Vietnamese War Bond; and (3) Safe Conduct Pass for NVA/VC deserters, distributed as part of the Republic of South Vietnam's psychological war effort.

13

GRENADES

The use of grenades for military purposes dates to the ancient Chinese. They were reported to have been used by the Roman army in 250 B.C. to combat elephant assaults. With the introduction of gun powder during the 13th century, the first use of explosive grenades was recorded. Prior to that time, grenades were used primarily as incendiary devices. Metal fire bombs affixed to rods and fired from muskets were used in the defense of the German city of Stettin in 1677. Under the reign of Louis XIV of France, grenadiers were considered elite troops. By the middle of the 18th century, all European armies had grenadier units. Grenades were first used on a massive scale during the 1905 Russo-Japanese War.

By 1913, the German army introduced the "potato masher" grenade. The British countered with their highly effective "Mills" bomb. The U.S. then developed its famous "pineapple" grenade. By World War II an extensive array of fragmentation and pyrotechnic grenades were in use by all armies of the world.

The Second Indochina War showcased nearly the entire history of grenades, ranging from simple fire bombs to highly sophisticated antitank grenades (the extent of the grenades actually used by PAVN and VC forces was mind boggling). World War II Japanese, Russian, Chinese and captured French grenades were used. These were augmented by a wide range of Soviet Bloc and ChiCom manufactured grenades, as well as modified, captured U.S. grenades and crude grenades, produced in jungle workshops. Using component parts supplied by allied Communist nations, crude factories, located in North Vietnam, were able to assemble an extensive number of grenades and other explosive devices starting in the late 1960s. Grenades were utilized in almost every type of guerrilla and infantry operation. Their reputation however was one of low explosive power and a (fortunate) general lack of reliability.

The two primary types of grenades used were the cast metal striker fuse variety and the stick grenade with wooden handle and a pull friction fuse. These were carried in two or four pocket pouches affixed to the web gear.

Viet Cong short handle (four to six inches), non-serrated head grenades with collar and wood plug and carrying pouch with shoulder strap.

PAVN antipersonnel rifle grenade with pull action fuse (left) and "toe popper" booby trap (foreground).

Two Viet Cong HEAT rifle grenades standing. PAVN HEAT rifle grenade foreground.

PAVN pull friction fuse, short handle grenade with large cylindrical head and carrying pouch with shoulder strap.

PAVN pull friction fuse, short handle grenade with serrated pineapple head and carrying pouch.

Short and two long handled ChiCom cylindrical head with collar grenades. Note that the center grenade has a threaded metal cap.

ChiCom pull friction fuse, long handle (8 to 9 inches) grenades with non serrated cylindrical head with collar.

U.S. M26 defensive hand grenade in "grenade drop" booby trap device.

Polish PGN-60 fin stabilized, point detonated HEAT rifle grenade (Foreground). French MDF polyvalent hand and rifle grenade (left) and Viet Cong copy of the French grenade (right) (background).

ChiCom copy of the POMZ-2M antipersonnel, fragmentation grenade/mine with serrated head.

ChiCom copy of the Soviet RKG3/3M/3T HEAT grenade. This grenade was used primarily by the VC to attack armored vehicles.

Polish F1/N60 antipersonnel, fragmentation rifle grenade. It is designed to be fired from a modified AK-47 assault rifle.

Viet Cong improvised fragmentation antipersonnel booby trap grenade with serrated concrete head.

Soviet POMZ-2M antipersonnel grenade/mine.

French brown jug defensive grenade (left), French offensive grenade (center) and Viet Cong copy (right).

Viet Cong improvised grenade made from the U.S. Mk3 grenade with a sheet metal liner added (left) and a U.S. antipersonnel cluster bomb converted to a grenade by the Viet Cong (right).

Soviet RKG3/3M/3T HEAT grenade (left) and practice version (right).

Three ChiCom grenades including a rare short handle grenade with long cylindrical head (center). Long handle grenade (left) has serrated fragmentation head. The long handle cylindrical grenade (right) has a threaded metal cap.

Viet Cong improvised pivot spoon grenade (left) and four PAVN striker fuse grenades (3 with serrated bodies and 1 with a smooth body, which was used extensively by North Korea).

(left to right) (1) Full size copy of the U.S. Mk2 defensive hand and rifle grenade; (2) assassin's copy of the U.S.Mk2; (3) improvised grenade with striker release fuse made from a French 50mm mortar bomb; and (4) Viet Cong improvised percussion fuse fragmentation grenade.

(left to right) (1) Viet Cong striker release fuse copy of French Model 35 grenade; (2) Viet Cong percussion fuse copy of the Japanese Type 98 grenade; (3) Japanese Type 98 grenade in leather pouch; (4) Japanese Type 98 grenade with Viet Cong striker release fuse; and (5) Viet Cong aluminum bodied copy of the Japanese Type 98 grenade with striker release fuse.

Viet Cong improvised aluminum body grenade (left) and ChiCom long handle cylindrical head with collar defensive grenade.

Two ChiCom short handle pineapple head grenades in belt pouch.

Viet Cong jungle workshop short handle grenades with wooden caps.

Viet Cong short and long handle cylindrical head grenades with wooden caps.

PAVN enlisted soldier with camouflage parachute material draped over his shoulders practicing throwing a satchel charge in a propaganda photograph. The charge is made from waterproof cloth strips tied around highly volatile potassium chlorate. A long handle stick grenade contains the detonator. These charges were used against bunkers and other forms of fortifications during assaults. Emering collection.

Various improvised jungle workshop grenades. Based on size alone, the large black bodied grenade in the background with a short handle is probably intended for use as a booby trap device.

(left to right) (1) Viet Cong improvised short handle grenade converted from a rifle grenade body; and (2) and (3) ChiCom long handle defensive grenades with serrated pineapple heads with a wooden (center) and metal (right) filler cap.

ChiCom long handle pineapple head grenade with threaded metal cap.

ChiCom long handle serrated pineapple head grenade with wooden filler cap.

(left to right) (1) Viet Cong striker release fuse tear gas grenade; (2) plastic body striker release fuse concussion grenade; (3) plastic body antipersonnel striker release fuse grenade; (4) Viet Cong improvised grenade; and (5) PAVN tear gas grenade with silver painted fuse.

ChiCom practice grenade.

Series of improvised jungle workshop grenades. The grenade at the far right foreground with yellow body markings is a PAVN pull friction fuse tear gas grenade.

Various ChiCom long and short handle stick grenades.

Three improvised jungle workshop grenades and a short handle pineapple head fragmentation grenade with carrying rope (laying down).

Soviet F1(ChiCom Type 1) serrated cast iron, defensive, antipersonnel, fragmentation grenade (left) and French 1935OF grenade (right) in canvas belt pouch.

(left to right) (1) ChiCom Type 1/M33 grenade with F1 fuse; (2) Soviet F1 antipersonnel hand grenade with ChiCom URZG fuse; (3) Soviet F1 grenade with shape plug; (4) Soviet RG42 (ChiCom Type 42) defensive grenade with ChiCom URZG fuse; and (5) Soviet RGD5 (ChiCom Type 59) defensive hand grenade with ChiCom URZG fuse.

French grenade (left) and Viet Cong friction fuse copy with short handle added (right).

Rare PAVN mini fragmentation grenade with ChiCom UZRG fuse (left) and PAVN version of the Soviet RGD5 defensive, sheet metal hand grenade.

Soviet F1 (ChiCom Type 1/M33) antipersonnel fragmentation grenade bodies adapted for use as booby traps.

Four ChiCom pineapple head grenades. The grenade at the far left has an aluminum body.

Captured Viet Cong jungle workshop quad grenade mold (center). U.S. Navy Museum, Washington, D.C.

Viet Cong jungle workshop dual fragmentation grenade mold.

14

MINES AND BOOBY TRAPS

The origin of mine warfare consisted of tunneling under otherwise impenetrable enemy positions and placing and detonating huge explosive charges. This exact tactic was used during the battle for Dien Bien Phu during May, 1954, when General Giap's Viet Minh forces perpetrated such an act upon the French. On May 6, 1954, the 102nd Infantry Regiment of the 308 PAVN Division, under the command of Huu Mai, after weeks of tunneling, detonated a huge cache of explosives under the French outpost Eliane 2, marking the end for its French occupiers.

During the latter part of World War I, both German and Allied armies began using land mines crudely fashioned from artillery shells. The importance of these tactics, however was not fully rec-

Danger sign marking a mine field in South Vietnam.

Soviet antitank mines TM-41 (right) and TMB-2 (left).

Viet Cong mortar shell mine (foreground) and Viet Cong improvised pressure firing device (rear).

Viet Cong antitank cast iron fragmentation mines. The larger device to the left was the standard fragmentation mine. The other three mines are improvised fragmentation devices.

Viet Cong cement turtle shaped mine.

ChiCom fixed directional antipersonnel fragmentation mine (DH-10). These mines were also mounted in trees and used to attack helicopters.

ognized until World War II, when the British army employed the tactic to bottle up the German forces in North Africa. The Soviet forces also used the tactic to delay the advance of the invading Nazi army until the deadly Russian winter arrived to fully thwart their advance.

Mines of all types have been used in every subsequent conflict of any note with Vietnam being no exception. A wide variety of devices were employed, including the reuse of captured ordnance.

Each mine is composed of six basic elements. These include: the pressure plug; fuse; case or body; booster charge; main charge;

Viet Cong directional antipersonnel fragmentation mine (MDH). Its improvised design is based on the U.S. Claymore mine.

Viet Cong antipersonnel "beer can" mine with cone shaped top (right rear); two improvised Viet Cong "shell case" mines in the foreground and a Viet Cong antipersonnel cylindrical mine (right rear) constructed in a Soviet shipping container. The fuse is concealed under the threaded red top.

Viet Cong hollow cone mine.

Viet Cong short cone shaped mine.

and the detonator. Mines can vary from the most sophisticated, shaped charges, such as Claymores to simple "booby" traps, frequently encountered during the Vietnam War. The body or case can be made from metal, paper, tar paper, asbestos, wood or an improvised, otherwise empty container such as a ration tin or beverage can.

The main types of mines are: (1) antitank; (2) antivehicular; (3) antipersonnel; (4) dual purpose (antipersonnel and antivehicular); (5) railroad mines; (6) beach mines; (7) river mines; and (8) improvised mines. The PAVN and VC placed a great emphasis on the use of improvised mines, which were often constructed in crude jungle workshops. Mines also provide an added psychological element in the conduct of warfare. The devastation caused by a mine, even a

Viet Cong cylindrical cement fragmentation (Soviet/ChiCom POMZ-2/2M type) mine (foreground) mounted on a wooden stake with pull type MUV-2 fuse projecting from the top and miniature fixed directional fragmentation mine (rear).

Viet Cong water mine.

Improvised antitank mine/booby trap made from two galvanized 7.62mm (AK-47) ammunition shipping containers wired together. The fuse well is drilled in the side of the bottom can. The device measures 14 inches long by 8 inches high.

Soviet TMD-B antitank wooden box mine (bottom) and two Soviet antipersonnel PMD-6 antipersonnel wooden blast mines with two piece wood casing.

simple "booby" trap can severely demoralize opposing forces. This was frequently the case in Vietnam, where antipersonnel mines caused horrendous casualties that did not always result in the death of the wounded individual.

Mines can be detonated by: (1) pressure; (2) pull; (3) pressure release; (4) tension; (5) delayed action; (6) vibration; (7) radio frequency induction; (8) magnetic induction; or (9) breaking or completing an electrical circuit.

The primary mines used during the Second Indochina War included:

ChiCom Mine #8 - A 9 inch diameter, 4 inch high cast iron mine using 5 pounds of TNT (a flammable solid chemically derived from toluene by a nitration process) as its main charge. It was usually painted black and required 30 pounds of pressure to detonate it.

Soviet Antitank Mine TM-41 - A 10 inch diameter, 5 inch high sheet metal mine, requiring 350 pounds of pressure for detonation.

It is often painted white or dark olive.

Soviet Antitank Mine TMB-2 - A 11 inch diameter, 6 inch high impregnated cardboard mine. Its main charge is 11 pounds of Amatol (a mixture of ammonium nitrate and TNT).

Viet Cong Mortar Shell Mine - A 4 inch diameter, 15 inch long cast iron mine, utilizing a modified mortar round. It weighs 13.5 pounds and uses 3.5 pounds of TNT as its main charge.

Viet Cong Fragmentation Mine - A 5 inch diameter, 9 inch long oval shaped cast iron mine. It is crisscrossed with serrations. It uses Melinite as its main charge.

Viet Cong Box Shaped Mine - An 8 inch diameter, 8 inch high cement mine. It is electronically detonated and uses TNT as its main charge.

Viet Cong Turtle Shaped Mine - Another cement mine, which is electronically fired. It is 5 inches in diameter and 9 inches long, weighing 13 pounds and using TNT as its main charge.

Two Soviet type PMD-6 antipersonnel wooden box mines (left and right) and a Soviet PMD-7ts antipersonnel wooden box mine (center). The lower casing of the PMD-7 is made from a solid wood block.

Booby trapped Czech T-21 Tarasnice 82mm recoiless gun round wooden shipping container.

Two Soviet PMD-6 antipersonnel wooden box mines and a Czechoslovakian PP-Mi-Sr "bounding" antipersonnel mine (center).

Soviet TMN-46 antitank mine. This mine is equipped with an anti lift device.

Viet Cong improvised DH-10 antipersonnel fragmentation mine.

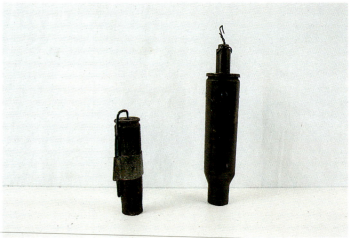

Viet Cong "toe popper" antipersonnel mine.

Viet Cong sheet metal "turtle" mine.

Viet Cong sheet metal "turtle" mine (variation).

VC guerrilla planting mine.

Viet Cong Directional Fragmentation Mine DH-10 - Intended for use to counter massed infantry attacks, this sheet metal mine was more often deployed as an antipersonnel booby trap. It consists of a 12 inch diameter, 2 inch wide body and bipods. It contains 420 to 450 cylindrical steel fragments, each 12mm in diameter. It has an effective range of more than 200 yards, yet can also be employed at much closer range in an antitank mode. The Viet Cong often used modified, captured U.S. Claymore mines in a similar role.

Viet Cong Cylindrical Fragmentation Mine - A 2 inch diameter, 6.5 inch long, homemade cast iron antipersonnel mine. The body is serrated for fragmentation effect. It uses a friction type ig-

niter and TNT as its main charge. A variation is a 7 inch wide, 22 inch long, electrically fired cement fragmentation mine.

Viet Cong Antipersonnel Mine - Constructed from a modified sheet metal grenade case, the 2 inch diameter, 6 inch long is ignited by pressure release on a spring mechanism.

Viet Cong Hollow Cone Mine - This sheet metal mine is 9 inches in diameter and 8 inches high. It weighs 15 pounds and utilizes a redundant ignition system.

Viet Cong Short Cone Shaped Mine - This electronically ignited, sheet metal mine contains 15 pounds of TNT. It is 11 inches in diameter and weighs 27 pounds. It has a handle attached to its side by two rivets.

Viet Cong improvised antipersonnel DH-10 type mine (rear). A small ChiCom antipersonnel wooden box mine is in the foreground (right) a "C" type battery pack and small ordnance fuse are on the left.

Five Viet Cong cylindrical cast iron fragmentation (stick grenade) mines with TNT filler, known as "Min" mines. Those with five serrated segments have a two to four second time delay friction igniter fuse; those with seven serrated segments have an instant fuse. These weapons have generally been classified as mines in official U.S. Army texts versus grenades, but with a weight of only 2.2 pounds and a short handle, those with the time delay fuse could theoretically be thrown like a stick grenade.

Viet Cong DH-10 mine with cutaway top and carrying handle (left), improvised antitank cement mine (center) and circular antipersonnel mine (right).

Viet Cong improvised sheet metal antitank mine.

Improvised antitank mine based on Soviet TM-46 sheet metal mine (left) and PAVN MDH-7 sheet metal antipersonnel mine (right).

ChiCom Type 61 blasting machine. This electrical impulse generator is capable of simultaneously firing up to 25 electric blasting caps wired in a series.

Viet Cong Water Mine - This black two compartment mine is fabricated from sheet metal. Painted black, it weighs in excess of 25 pounds and utilizes TNT as its main charge. The lower compartment is hollow and is used to keep the mine buoyant.

It would be inappropriate to close this chapter without mentioning one of the most insidious booby traps employed by the enemy during the Second Indochina War.

Although non-explosive in nature, the Caltrop spike or "punji" stick was capable of piercing the standard combat shoe and inflicting severe pain and injury. The device was fashioned from 2 to 12 inch metal spikes and/or sharpened raw or fire hardened bamboo sticks dipped in human or animal excrement to introduce infection to the wound. The Viet Cong became expert in deployment and use of these traps. The injuries caused almost always necessitated a medical evacuation. The devices were often employed with multi-directional spikes to make extraction even more difficult.

From January, 1965 through June, 1970, booby traps, particularly mines, accounted for 11 percent of all combat deaths and 17 percent of all combat injuries. They were an inexpensive and very effective means of combat and their psychological effect on troop morale could not be ignored.

For the collector of inert examples of mine warfare, examples of mines used during the Vietnam War seem endless, given the proclivity of the enemy forces to use improvised devices. The mines mentioned above are but a brief sampling of a nearly limitless variety of such devices employed by the enemy forces.

Soviet PM-1 hand operated, low tension, electrical generator blasting machine, patterned after the German WW II Field Exploder (1948). It is capable of simultaneously detonating up to 100 electric blasting caps.

Soviet PM-2 blasting machine (1953).

ChiCom LA 2 B blasting machine (copy of the Soviet PM-2) capable of simultaneously firing up to 10 electric blasting caps.

Soviet KPM-1D blasting machine (1957).

Soviet PM-1 blasting machine variation (1949).

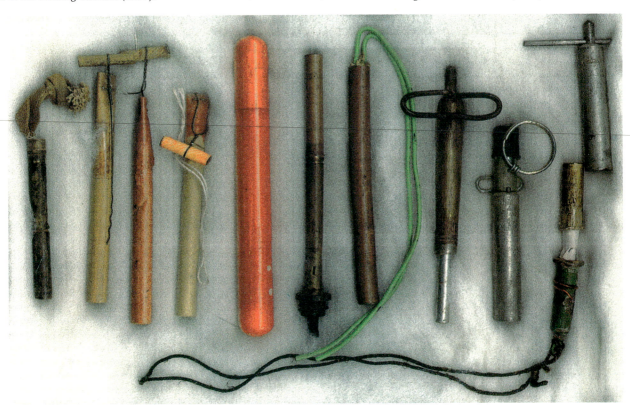

(left to right): (1) through (4) ChiCom/Soviet pull type fuse igniters; (5) Soviet MV-5 pressure fuse; (6)Soviet tension firing device, M-VPF (copy of the U.S M-3); (7) PAVN/VC improvised electric blasting cap; (8) Soviet MUV pull trigger device for the POMZ-2 and PMD-6 antipersonnel mines; (9) French three to four second delay, pull-safety fuse with pull string; (10 - bottom) Soviet and ChiCom pull fuse VPF; and (11 - top) PAVN/VC pressure or pull activated firing device.

"Punji" stake trap uncovered by U.S. Army patrol. U.S. Army photo.

Four variations of steel caltrops and a booby trapped food tin.

The infamous bamboo "punji" stakes, which were often dipped in human or animal excrement in order to spread infection to the victim.

15

ORDNANCE

The vast majority of ordnance used during the Second Indochina War was of Soviet or ChiCom origin. In many cases the Chinese calligraphy or Soviet Cyrillic lettering is clearly visible on the individual rocket or shell casing, verifying its source of origin. Nearly every item of Soviet ordnance was copied by the Chinese and so each piece of Soviet ordnance usually had its ChiCom counterpart.

The NLF was particularly adept at converting spent ordnance casings into "booby" traps or other forms of homemade ordnance. They would at times also fire the ordnance from field expedient devices. While severely limiting the accuracy of the ordnance, they were still able to make effective use of it, especially when firing randomly into a base or village.

60mm mortar bomb.

82mm mortar bomb.

Section of 122mm rocket shell which reads: "From your Friends the Soviets" in Cyrillic and Vietnamese.

Soviet 240mm rocket.

RPG-7 rocket.

Bangalore torpedo (variation).

VC improvised rocket made partly from an RPG-2 round. The jacket is made from galvanized tin hammered and soldered into shape. The stabilizing fins are an improvisation.

ChiCom 76mm recoiless rifle round.

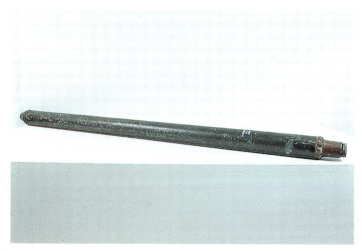

Bangalore torpedo used for breaching wire.

Three ChiCom 60mm (tall) and a French 50mm (short) mortar bomb.

Four 60mm mortar bombs.

Two 82mm mortar bombs.

ChiCom Type 50 HEAT Grenades fired from the RPG-2 (right and left) and an 82mm antipersonnel mortar bomb (center).

Two 82mm mortar bombs (the one on the left still has its shipping plug).

ChiCom 87mm rocket for the Type 51 antitank launcher with 120mm mortar fuse (left) and the 87mm rocket fuse (right).

ChiCom 107mm spin stabilized "true" rocket shell.

Jungle workshop improvised round.

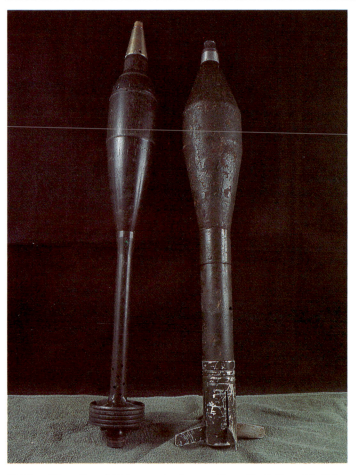

ChiCom 87mm Type 241 heat rocket for recoiless rifle (left) and B-50 improvised round (right).

"Fish Hook" stabilizing fin assembly for the ChiCom B-41 rocket. The tail is opened after launching by centrifugal force and gives the round a clockwise spin.

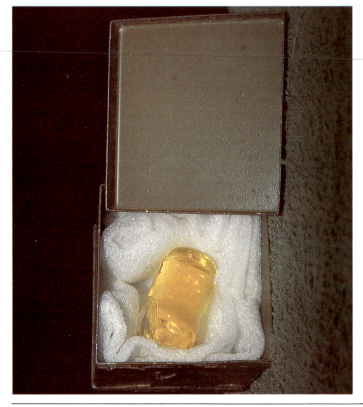

LEFT: Acid ampule for 82mm chemical delay mortar bomb fuse.

Launching cartridges (motor elements) for RPG-2 rocket.

ChiCom fin stabilized 75mm HEAT Cartridge (Type 56) and carrying container.

ChiCom 82mm HEAT Recoiless Rifle round and carrying case.

ChiCom Bangalore torpedo with canvas carrying strap. This weapon, which is approximately 40 inches in length, was used as an antitank weapon. The charge is ignited by a simple stick handle grenade at the base of the tube.

16

FLAGS

In addition to the standard PAVN and National Liberation Front flags, which came in many sizes, the North Vietnamese made use of a wide variety of pennants, honor flags (celebrating victories both real and imagined) and political banners. Many unit (battle) flags have been observed on formal occasions adorned with medals (both Orders and Decorations) awarded to the unit. Nearly all such awards can be earned individually or collectively. Unit flags often contain slogans such as "Quyet Thang" (Resolved to Win), "Chien Thang" (Victory) or "Giai Phong" (Liberation). They may also contain the unit number, such as "308B." Flags apparently played an important role in the overall morale of the troops and were considered important possessions of the unit.

Cotton NLF version of the Ap Bac Honor Flag (30 inches by 40 Inches). Photo courtesy of Gerry Schooler.

Socialist Republic of Vietnam National Flag captured July 1967 from the 126 Naval Sapper Group in the vicinity of the Cua Viet River.

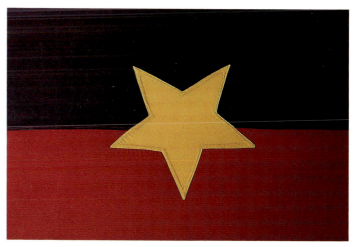

National Liberation Front Flag.

Satin Ap Bac honor flag commemorating the NLF's stunning and prophetic victory over the South Vietnamese Army on January 2, 1963.

Portion of a typical political banner.

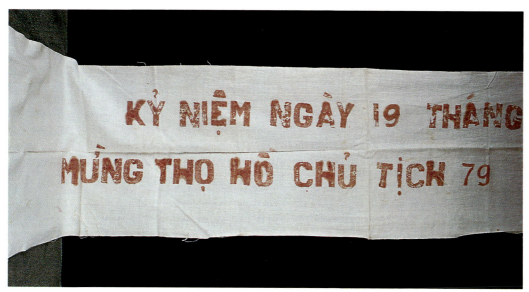

Homemade victory commemorative banner.

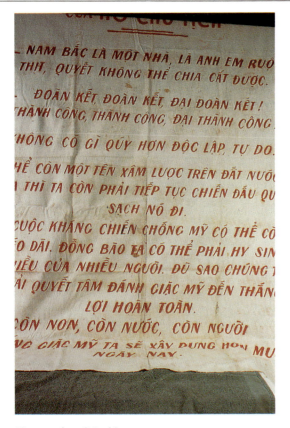

Homemade political banner.

NLF "Peoples' Revolutionary Volunteer Unit of South Vietnam" honor flag.

Communist Party flag.

PAVN unit emulation pennant.

SRV pennant.

Cong An (security police) pennant.

Ho Chi Minh Labor Youth Group honor flag.

Satin PLAF uni: flag.

SRV unit pennant.

NLF pennant.

PAVN emulation flag for Heroes Who Kill Americans (Dung Si Diet My).

PAVN rotating emulation flag for excellent units.

Rotating emulation flag for Labor Youth Unit 37 with embroidered Labor Youth Badge in center.

559th Transportation Group unit flag for operations inside South Vietnam. This secret group, which operated the Ho Chi Minh Trail, maintained the pretext that the National Liberation Front operations in South Vietnam were completely separate from Hanoi's control or influence. This flag would have been part of such ruse.

Three Firsts (First in Combat, First in Training & First in Discipline) emulation flag.

PAVN 308 Division Spring Offensive (1972) unit flag with the 1972 Nguyen Hue Offensive campaign badge embroidered in the center.

PAVN Determined to Win (Quyet Thang) flag.

PAVN Naval Infantry (Thuy-Quan Luc-Chien) emulation flag for Heroes Who Kill Americans (Dung-Si Diet My).

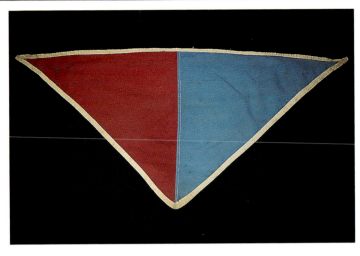

NLF cotton pennant.

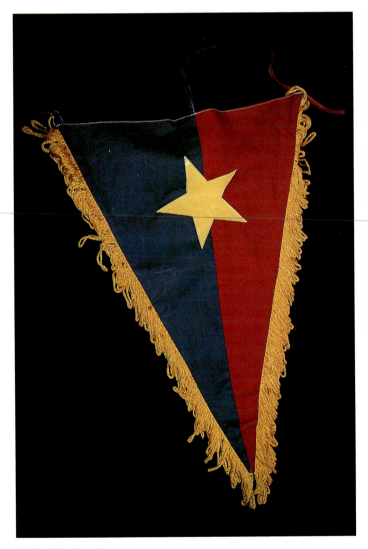

NLF embroidered pennant.

PAVN cloth pennant.

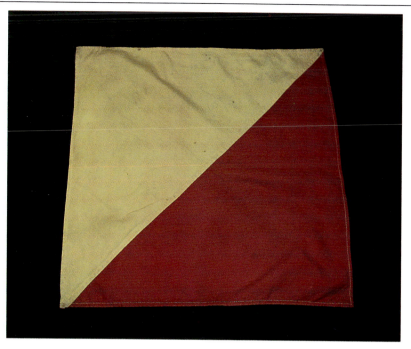

PAVN cloth square.

PAVN troops massed beneath political banners in Hanoi during the Second Indochina War.

17

MISCELLANEOUS EQUIPMENT

The following items of NVA/VC field gear did not fit any of the previously specified categories and are therefore grouped together here as miscellaneous equipment.

Flamethrower LPO-50 - The Soviet Manpack Flamethrower LP-50 consists of three cylindrical fuel tanks mounted on the carrier's back and a rifle-like launcher fitted with a folding bipod. It replaces the ROKS-2 and ROKS-3 portable flamethrowers. The flame gun is capable of a range of 70 meters with thickened fuel. The entire assembly weighs approximately 51 pounds when full. No separate compressed air bottle is required, since each tank is fitted with a pressurizing charge.

Artillery Range Finder - This device was used to visually adjust fire by an artillery battery commander. It is archaic considering

Soviet Manpack Flamethrower LPO-50.

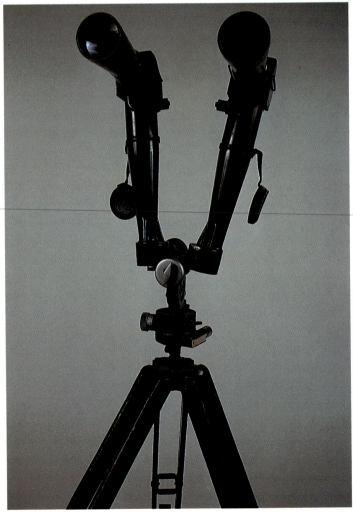

Obsolete Soviet "trench style" Visual Artillery Range Finder with "rabbit ear" binoculars.

just how obsolete visual fire control was even at the time of the Second Indochina War.

Antiaircraft Range Finder - This was also an obsolete visual range finder employed during the Second Indochina War.

Soviet IMP Mine Detector - This replacement for the UMIV-1 portable mine detector is a transistorized manpack capable of operating on land and under water. The detector head is housed in a plastic cylinder mounted on a four-part collapsible aluminum pole. The controls and batteries are located in a small rectangular box, comprised of the tone-regulator set over the tuning control, which uses five transistors, powered by four 1.5 volt dry batteries. The headphones and detector-head are connected directly by cable to the control box. The entire assembly comes with a light weight metal storage container.

Soviet Military Compass - This is a lightweight Soviet compass with an aluminum plate inside the top of the case to aid in reading the compass and for use as a signaling device.

Other miscellaneous militaria items included in this Section are a small ChiCom compass and a ChiCom copy of the Japanese 27mm double barrel signal pistol.

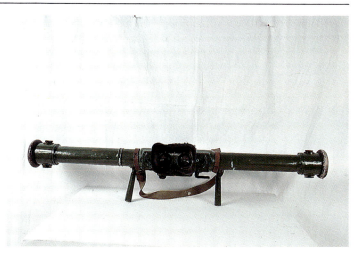

Obsolete Soviet Visual Antiaircraft Range Finder.

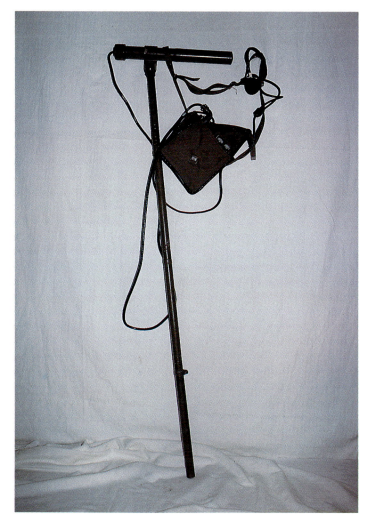

Soviet IMP Manpack Mine Detector.

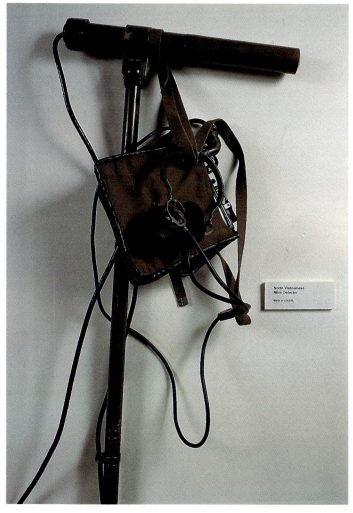

Soviet IMP Mine Detector, close-up of battery pack and detection monitor.

133

Soviet Compass with the four major compass points indicated by Cyrillic vowels. The smooth metal plate inside the aluminum top cover of the compass acts as both an aid in reading the compass as well as a signaling mirror. The dial is divided into 6,000 mils instead of the usual 6,400 mils used on a U.S. compass. Emering collection.

ChiCom copy of a Japanese 27mm double barrel, two shot signal pistol. The barrel release knob adjusts to three positions: right, left and safety.

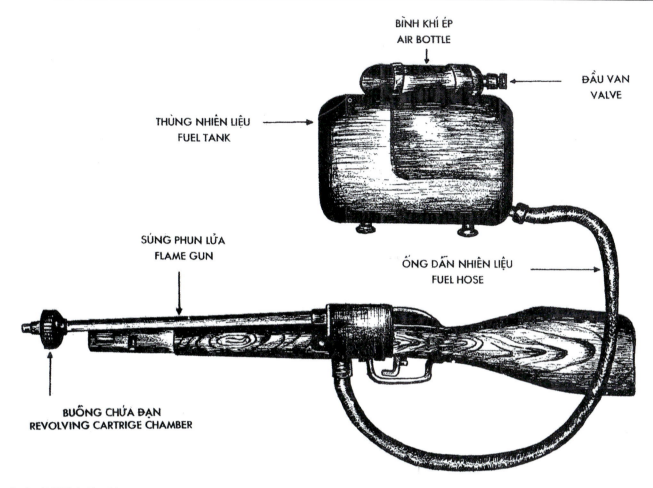

BÌNH KHÍ ÉP
AIR BOTTLE

ĐẦU VAN
VALVE

THÙNG NHIÊN LIỆU
FUEL TANK

SÚNG PHUN LỬA
FLAME GUN

ỐNG DẪN NHIÊN LIỆU
FUEL HOSE

BUỒNG CHỨA ĐẠN
REVOLVING CARTRIGE CHAMBER

Soviet ROKS-3 Shoulder Carried Flame Thrower. The weapon fires a stream of flaming liquid when the trigger is pulled. It contains sufficient fuel in its single tank compartment for six to seven blasts. Chien Cu Part I.

Captured main force Viet Cong gear with: (1) two pocket button cuff shirt; (2) Ho Chi Minh sandals; (3) boonie cap with camouflage patches sewn on to it; (4) canteen with Bakelite top, padded cover and shoulder strap; (5) porcelain rice bowl; and (6) glass trail lantern (in bowl). Peter Aitken collection.

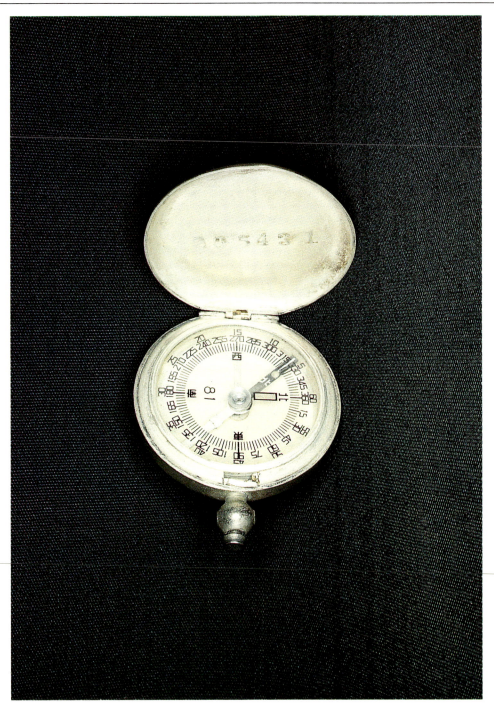

ChiCom compass.

18

FACES OF THE ENEMY

This chapter presents a series of captured enemy photographs from the F. C. Brown Collection at the National Vietnam Veterans Art Museum in Chicago. They represent not only an important militaria collectible, themselves, but also depict the wide variety of dress and equipment used by the enemy forces fighting in the South, including "fillers" provided to the National Liberation Front by the PAVN and infiltrated into South Vietnam along the Ho Chi Minh Trail. The photographs are also included to humanize the enemy for those who have not had the first hand experience of the Vietnam War.

Viet Cong main force officer in bamboo helmet and wearing Ho Chi MInh sandals.

Viet Cong soldier with captured U.S. .30 caliber Browning Automatic Rifle M1918A2.

Viet Cong squad carrying a ChiCom Type 52 75mm recoiless rifle.

Viet Cong soldier with M-1/M-2 carbine rifle.

Viet Cong main force officer with ChiCom Type 54 holster and padded canteen cover.

Viet Cong soldier with fixed stock ChiCom Type 56 version of the AK-47 assault rifle. He is also wearing a soft boonie cap and an AK-47 chest pouch.

Weavers in a jungle workshop.

Jungle workshop scene.

Viet Cong soldier with Ho Chi Minh sandals dismantling live U.S. aerial bomb in a jungle workshop.

Jungle workers dismantling live U.S. aerial ordnance.

*Two VC guerrillas in black paja-
mas and Ho Chi Minh sandals.*

Three VC nurses (circa 1967).

Viet Cong main force officer, most likely a PAVN filler (circa 1967). He is wearing a U.S. made web belt with a ChiCom Type 54 holster and Ho Chi Minh sandals.

VC Squad in a "water taxi."

PAVN non-commissioned officer.

Main force Viet Cong officer with ChiCom Type 54 holster, utility pouch and padded canteen cover on his web belt. Note the display of the "hero" pen in his left breast pocket.

Two main force Viet Cong soldiers.

Female Viet Cong guerrilla with checkered scarf/towel.

Female VC guerrilla with captured U.S. caliber .50 machine gun M2.

Viet Cong guerrilla in black pajama uniform.

PAVN Signal Corps non-commissioned officer.

Viet Cong squad poses with mixture of uniforms and weapons, including a ChiCom Type 56 copy of the Soviet RPD LMG.

Viet Cong mortar squad poses for a propaganda film in III Corps (circa 1966).

Viet Cong officer (left, with pens) and three soldiers pose for a photo opportunity.

"Nothing is more precious than independence and liberty."

- Chu Tich (Chairman) Ho Chi Minh

"We fight and win not because we are endowed with steel skin or copper bones but....because we are Vietnamese..."

- Le Duan, General Secretary of the Vietnamese Communist Party

"The enemy will pass slowly from the offensive to the defensive. The blitzkrieg will transform itself into a war of long duration. Thus, the enemy will be caught in a dilemma: he has to drag out the war in order to win it and does not possess, on the other hand, the psychological and political means to fight a long drawn-out war."

- Senior PAVN General Vo Nguyen Giap, original Commander-in-Chief of the Viet Minh and subsequently the PAVN.

"War is the highest, most comprehensive test of a nation and its social system. War is a contest that not only tests the skill and strategy of two adversaries, but also their strength and will. Victory goes to the side which has the correct military strategy, which makes the best use of the art of military science and which most successfully limits the war making capacity of its adversary."

- Senior PAVN General Van Tien Dung, successor to General Giap as Commander-in-Chief of the PAVN

VALUE GUIDE

This Value Guide is provided as a further reference to the reader. In many cases, the number of permutations and the actual condition of the weapon or equipment will significantly impact the item's value. Also, in terms of the U.S. market for Vietnam War militaria, the ability to identify a particular item as a "war era" piece will enhance its value. In some cases, the collector will be able to locate a date stamp or some other indication of the items date of manufacture. A prime example of the variation in value can be noted with regard to the sun helmet. Prior to standardization of PAVN headgear in 1982, Coastal Defense personnel often wore white sun helmets. These white sun helmets are not only rare, but are also almost always "war" vintage. Their price is more than ten times the value of the standard khaki sun helmet, which continues to be manufactured to this day.

Since all of these various considerations are too numerous to cover in a guide such as this, the reader must consider it to be of a general nature. In other cases, the fact that items were unique or "homemade" creates a significant impact on the value of the underlying piece. In certain cases, the rarity of the item requires that its value be established via the auction process much like the value of a rare painting. I have tried therefore to present broad value ranges for each weapon or piece of equipment. The collector should therefore be advised to use this only as a general guide and not a definitive statement as to the value of any particular piece being considered for purchase. In certain cases, where a piece exists in such limited numbers that a general value cannot be provided, the indication "rare" or "R" has been substituted for a general value range.

I would also like to thank Robert Pucci of the Wisconsin Military Armaments Museum in Janesville, Wisconsin for his assistance in helping to place values on the inert weapons depicted in Chapters 1 and 2 of this work.

The position codes are as follows:

T	Top
B	Bottom
C	Center
L	Left
R	Right
TL	Top Left
TO	Top Center
TR	Top Right
TRC	Top Right Center
TLC	Top Left Center
BL	Bottom Left
BR	Bottom Right
CL	Center Left
CR	Center Right
CT	Center Top
CB	Center Bottom

The Value Codes are as follows:

A	Under $50.00
B	$50 up to $100
C	$101 up to $250
D	$251 up to $500
E	$501 up to $1,000
F	$1,000 up to $2,500
G	Over $2,500
R	Too Rare to Value

Page	Position	Value Code			
			32	BL	A
			32	BR	C
9	BL	C	33	TL	B
9	BR	R	33	TR	A
10	TL	R	33	BL	B
10	TR	R	33	BR	C
10	CL	R	34	TL	B
10	CR	D	34	TR	D
10	BL	D	34	BL	C
10	BR	R	34	BR	B
11	TL	R	35	T	B
11	TR	R	35	BL	B
11	BL	C	35	BR	D
11	BR	D	36	T	C
13	TL	D	36	BL	D
13	TR	D	36	BR	C
14	TL	C	37	TL	C
14	TR	C	37	TR	B
14	CR	E	37	CR	B
14	BL	C	37	B	B
14	BR	E	38	TL	C
15	TL	D	38	TR	B
15	TR	D	38	BL	D,D
16	TR	E	38	BR	A
17	TL	E	46	TL	B
17	TR	E	46	TR	B
17	CL	E	46	CTL	B,B
17	CR	E	46	CTR	B,B
17	BL	F	46	CBL	B
17	BR	E	46	CBR	B
19	TL	E	46	BL	C
19	TR	E	46	BR	B
19	B	E	47	L	C
21	TL	F	47	R	C
21	TR	F	48	TL	C
21	CL	G	48	TR	B
21	CR	G	48	BL	B
21	BL	R	48	BR	B
21	BR	F	49	TL	C
23	TL	F	49	TR	B
23	TR	F	49	BR	B
23	B	F	49	BL	B
24	TL	D	50	TL	C
24	BL	E	50	TR	B
24	BR	F	50	BL	B
25	BL	E	50	BR	B
25	BR	F	51	TL	B
29	BL	B	51	TR	C
29	BR	A	51	BL	C
30	TL	B	51	BR	B
30	TR	B	52	TL	C
30	BL	C	52	TR	C
30	BR	B	52	BL	C
32	TL	B	52	BR	C
32	CL	B	53	TL	B

53	TR	C		74	BL	A,A
53	BL	B		74	BR	B
53	BR	B		75	T	B
54	TL	B		75	B	B
54	TR	B		76	BL	B
55	L	C		76	BR	B
56	BL	B		77	TL	B
56	BR	A		77	TR	B
57	TL	A		77	BL	C
57	CL	A		77	BR	B
57	CR	B		78	BL	B
57	B	B		78	BR	C**
58	L	B		79	TL	A,B
58	R	D		79	TR	B
60	TL	F		79	CL	B
60	TR	D		79	CR	B
60	BL	D		79	BL	A
60	BR	C		79	BR	C
61	TL	B		80		C
61	TR	C		81		C
61	BL	B		82	TL	C
61	BR	C		82	TR	A
62	TL	D		82	CL	B
62	BR	C		82	BL	B
63	BL	B		82	BR	B
63	BR	B		83		B
64	TL	B		84		B
64	TR	B		85	TR	R
64	BL	B		85	CL	B
65	TL	B		85	CR	B
65	TR	B		85	BL	C
65	CL	B		85	BR	B
65	CR	B		86	TL	B
65	B	B		86	TR	C
66	TL	B		86	CL	B
66	TR	B		86	CR	B
66	BL	B		86	BL	B
66	BR	B		86	BR	C
67	L	B		87	TR	C
67	R	C		87	CL	C
68	TL	B		87	CR	C
68	TR	B		87	BL	C
68	BL	B		87	BR	C
68	BR	B		88	TL	C
69	TR	B		88	TR	B
69	BL	A		88	BL	C
69	BR	A		88	BR	B
71	BL	A		89	TL	B
71	BR	C		89	TR	C
72	TL	A		89	BL	B
72	TR	B		89	BR	B
74	TL	A		90	BL	B,B,B,B
74	TR	C**		90	BR	B,C,B,B
74	CL	C**		91	TL	B
74	CR	R*		91	TR	C

91	BL	B,B	104	BR	B,B,B
91	BR	C	105	TL	B
92	TL	B	105	TR	B
92	TR	B	105	BL	B,B,BB,B
92	BL	B	105	BR	B
92	BR	C	106	TL	B (all)
93	TL	B	106	TR	B (all)
93	TR	B	106	CL	B,B,B,B
93	BL	B	106	CR	B,B
93	BR	B	106	BL	B,B,B,B,B
94	TL	B,B,B	106	BR	B,B
94	TR	B,C,C	107	TL	C,B
94	CL	C**	107	TR	B,B
94	CR	C**	107	CL	B,B,B,B
94	BL	C**	107	CR	C
94	BR	A,A,A,A	107	B	C
95	TL	A,B,C,B,B,B,B,B	108	TL	C
95	TR	B	108	TR	B,B
95	BL	C	108	BL	C
95	BR	C	108	BR	C,C,C,C
96	TL	B	109	TL	C
96	TR	C	109	TR	C
96	CL	B,B, B	109	BL	C
96	CR	B	109	BR	B,B,B,B
96	BL	C	110	TL	C
96	BR	B,B	110	TR	C
97	T	A	110	BL	B,C
97	C	A	110	BR	C
97	B	A	111	TL	B
98	BL	B	111	TR	B,B,B
98	BR	B,B	111	BL	B,B,B
99	TL	B,B,B	111	BR	B
99	TR	C**	112	TL	B,B,B
99	CL	C	112	TR	C
99	BL	B,B,B	112	CL	B
99	BR	B	112	CR	B,B
100	TL	B	112	BL	C
100	TR	B,B,B	112	BR	C
100	BL	B	114	TL	B,B,B,B
100,	BR	B	114	TR	B,B,B,B,B
101	TL	B	114	CL	B,B,B
101	TC	B	114	CR	B
101	TR	B	114	BL	B,B
101	BL	B,B,B	114	BR	C
101	BR	B,B	115	TR	C
102	TL	B,B	115	BL	C
102	TR	B,C,B	115	BR	C
102	BL	B,B,B,B,B	116	TL	C
102	BR	B,C,B,B	116	BL	C
103	TL	B,B,B,B,B	116	BR	C**
103	TR	B,B	117	BL	B,B,B,B,B
103	CL	B,B	117	BR	B
103	CR	B,B,B,B	118	TL	B
103	B	B,B,B,B	118	R	B
104	BL	B,B,B,B,B,B,B,B	118	CL	C

118	BL	C	126	BL	B
119	TL	C	126	BR	B
119	TLC	B	127	TL	B
119	TRC	C	127	TR	B
119	TR	C	127	BL	C
119	BL	B	127	BR	C
119	BR	B,B,B,B	128	TL	B
120	T	B,B,B,B	128	TR	B
120	BL	B,B	128	BL	C
120	BR	B,B,B	128	BR	C
121	TL	B,B	129	TL	C
121	TR	B,C,B	129	TR	C
121	BL	C	129	CL	C
121	BR	B,B	129	BL	D
122	TL	B,B	129	BR	C
122	TR	B	130	TL	D
122	BL	B	130	TR	B
123	TL	B	130	BL	C
123	TR	B	130	BR	B
123	BL	B	131	T	A
123	BR	C	132	L	D
124	D		132	R	C
125	TL	C	133	TR	C
125	TR	C	133	BL	D
125	CL	E	134	T	C
125	CR	B	134	B	C
125	B	C	135	T	D
126	TL	B	135	B	D**
126	TR	C	136		C

*It may be illegal in some jurisdications to possess the contents of these vials. Please consult local laws and regulations before attempting to acquire such collectibles.

**Entire grouping.

All photos displayed in Chapter 18 are valued between $25.00 and $50.00 each.

BIBLIOGRAPHY

Chien Cu (War Material) Part I & Part II:Republic of Vietnam Armed Forces, 1964.

Conboy, Ken, Bowra, Ken and McCouaig, Simon: *The NVA and Viet Cong.*
London, England: Osprey Publishing, Ltd., 1991.

Emering, Edward J.: *Orders, Decorations and Badges of the Socialist Republic of Vietnam and the National Front for the Liberation of South Vietnam.* Atglen, Pa: Schiffer Publishing Ltd., 1996.

Grenades and Pyrotechnic Signals Field Manual 23-30. Washington, D.C.: Department of the Army, December, 1969.

Handbook of the North Vietnamese Armed Forces. Washington, D.C.: Department of the Army, Pamphlet No. 30-53, December 6, 1961.

Janes Infantry Weapons 1976, 2nd Edition. London, England, 1976.

Katcher, Philip and Chappell, Mike: *Armies of the Vietnam War 1962-75.*
London, England: Osprey Publishing, Ltd., 1980.

Kutler, Stanley I.: *Encyclopedia of the Vietnam War.* New York, New York: MacMillan Library Reference, 1996.

Lanning, Michael Lee and Cragg, Dan: *Inside the VC and the NVA.* New York, NY: Ballantine Books, 1992.

Lulling, Darrel R.: *Communist Militara of the Vietnam War Revised Edition.* Tulsa, OK: M.C.N. Press, 1980.

Mathiesen, Chris: *VC - NVA Ordnance.* Undated monograph.

McLean, Donald B.: *Guide to Viet Cong Ammunition.* Forest Grove, OR: Normount Technical Publications, 1971.

Plaster, John L.: *SOG The Secret Wars of America's Commandos in Vietnam.* New York, NY: Simon & Schuster, 1997.

Rideout, Granville N.: *The CHICOM Series.* Ashburnham, MA: Yankee Publishing Company, 1971.

Russell, Lee E. and Chappell, Mike: *Armies of the Vietnam War 2.* London, England: Osprey Publishing, Ltd., 1983.

Smith, Graham: *Military Small Arms 300 Years of Soldiers' Firearms.* London, England: Salamander Books Limited, 1994.

Soviet Mine Warfare Equipment Technical Manual TM5-223A.. Washington, D.C.: Department of the Army, August, 1951.

Truong Nhu Tang with David Chanoff and Doan Van Toai: *A Viet Cong Memoir.*
New York, New York: Vantage Books division of Random House, 1985.

VC/NVA Employment of Mines & Booby Traps. San Francisco, CA: Headquarters U. S. Military Assistance Command, Vietnam, Office of the Assistant Chief of Staff, Intelligence, Technical Intelligence Branch, Combined Intelligence Center, Vietnam, 1967.

VC/NVA Rocket Artillery. San Francisco, CA: Headquarters U.S. Military Assistance Command, Vietnam Office of the Assistant Chief of Staff, Intelligence, 1967.

Vietcong Mines & Boobytraps. El Dorado, AK: Delta Press, undated.

Weapons and Equipment Recognition Guide: Southeast Asia. Washington, D.C.: Department of the Army, Pamphlet 381-10, March, 1969.

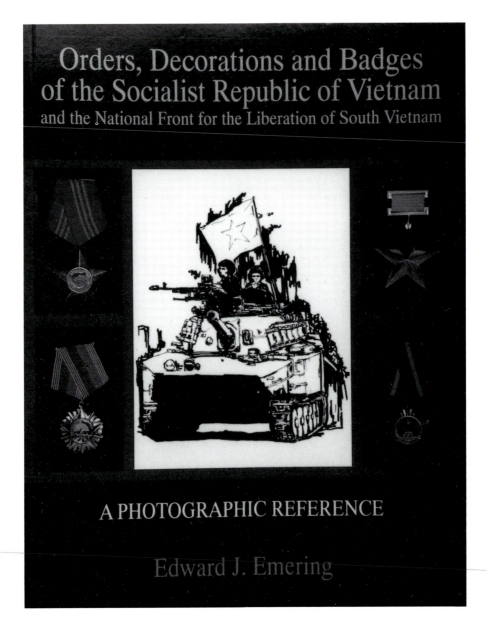

ORDERS, DECORATIONS AND BADGES
OF THE SOCIALIST REPUBLIC OF VIETNAM
and the National Front for the Liberation of South Vietnam

Edward J. Emering

The Orders and Decorations of the "enemy" during the Vietnam War have remained shrouded in mystery for many years. References to them are scarce and interrogations of captives during the war often led to the proliferation of misinformation concerning them. Covered are those Orders and Decorations now considered official by the SRV, as well as many of the obsolete awards bestowed by the DRV and the NLF. Includes value guide.

Size: 8 1/2" x 11",
190 color and b/w photographs, line drawings
96 pages, soft cover
ISBN: 0-7643-0143-8 $24.95